Future-Ready Leadership

Strategies for the Fourth Industrial Revolution

Chris R. Groscurth, PhD

PRAEGER™

An Imprint of ABC-CLIO, LLC

Santa Barbara, California • Denver, Colorado

Library of Congress Cataloging-in-Publication Data

Names: Groscurth, Chris R., author.
Title: Future-ready leadership : strategies for the fourth industrial revolution / Chris R. Groscurth, PhD.
Description: Santa Barbara, California : Praeger, [2018] | Includes bibliographical references and index.
Identifiers: LCCN 2018015177 (print) | LCCN 2018015999 (ebook) | ISBN 9781440865237 (e-book) | ISBN 9781440865220 (hard copy : alk. paper)
Subjects: LCSH: Leadership. | Strategic planning.
Classification: LCC HD57.7 (ebook) | LCC HD57.7 .G7654 2019 (print) | DDC 658.4/092—dc23
LC record available at https://lccn.loc.gov/2018015177

ISBN: 978-1-4408-6522-0 (print)
      978-1-4408-6523-7 (ebook)

22 21 20 19 18   1 2 3 4 5

This book is also available as an eBook.

Praeger
An Imprint of ABC-CLIO, LLC

ABC-CLIO, LLC
130 Cremona Drive, P.O. Box 1911
Santa Barbara, California 93116-1911
www.abc-clio.com

This book is printed on acid-free paper ∞

Manufactured in the United States of America

Contents

Acknowledgments

I am grateful to the following people for their inspiration, support, and encouragement while I was researching and writing this book.

First, to my clients Larry Kleinman, Joe Pellegrini, and Gen Coleman for their partnership over the past few years and for taking bold actions to prepare tomorrow's leaders today.

Second, to my friends and colleagues Dipak Sundaram, Camille Patrick, William Busch III, Sarah Houle, Jillian Anderson, Kelsey Seabolt, Carole and Dave Schwinn, and Rox Pals for their curiosity, creativity, and thoughtful engagement with me about the future of work, leadership, and change management.

From a production standpoint, this project wouldn't have been possible without a tremendous Praeger/ABC-CLIO team. I am particularly grateful to Hilary Claggett for her belief in this project. And to Rachel Kaboff (rkaboff .com) for being an amazing gig economy partner, providing design support on many different iterations of the figures in the book.

Finally, to my friends and family who supported me while I was making the future of my work and life a reality—thank you. To Beverly Severance and Shannon Groscurth for always nurturing my love of language. To Lynn Jacques Smith for introducing me to the life and works of Blessed Solanus Casey. To Dr. Joe "Fishstix" Fanning for his thought partnership, marketing wisdom, and friendship. And lastly, to the love of my life, Robin VanDenabeele. Thank you for your encouragement, patience, and balance.

CHAPTER ONE

Leadership for the Fourth Industrial Revolution

I am enjoying a Grande Americano at the 15th St. Starbucks in Omaha, Nebraska, when Howard Schultz, executive chairman of Starbucks, sits down next to me. "Chris," Schultz says, "my head of business strategy has convinced me to reduce Starbucks' physical footprint and invest in mixed-reality cafes that are fully automated by robots. It's called robotic process automation or RPA. The artificial intelligence (AI) capabilities that these bots have is remarkable. They'll learn when, where, and what our customers love to buy through data collected via a small wearable device. These bots will take up a fraction of the space of a brick-and-mortar store, and the store redesign will offer customers a completely customizable mixed-reality experience. Think of it this way, instead of having coffee with me in these old boring brown leather chairs, we could have coffee together on a beach in the Caribbean, on your private jet, or on the moon for that matter!"

"Wow, Mr. Schultz. That sounds like a very exciting change in the customer experience."

"Yes, it has great potential. But listen, the reason why I stopped by to share this with you is because I know that you help leaders get their organizations ready for the future. So, my question for you is, with all of the promise that our mixed-reality and artificial intelligence model offers, what about our people? Is this fair to them?"

At that moment Schultz and the cafe in Omaha melt away in front of my eyes. I'm back in a conference room at 111 S. Wacker Dr. in Chicago. Carefully, I remove my Magic Leap photonic glasses and I set them on the table. These are the tools that have allowed me to experience this virtual reality encounter. Jim is sitting across from me in a navy suit. He is a partner from

Deloitte's people and change practice. I take a deep breath. "Well," asks Jim, "how would you advise Mr. Schultz under these circumstances?"

This isn't a dream about a job interview gone awry. This is a glimpse of the future of work. Everything in this scenario is technologically possible today: from the smart, connected AI-powered coffee shop, to the use of augmented and virtual reality technologies for interviewing and training. Although there are plenty of unknowns about artificial intelligence and the fourth industrial revolution, one certainty surrounds the future of work: *future-ready leadership will be more important than ever before.* If you want to successfully lead your organization forward into the future, then you need to understand the new rules of leading in the fourth industrial revolution (4IR).

In this book I explain leadership's role in building organizations of the future. It's a book about how to prepare the people in your organization to work toward a shared purpose and a shared future. At their core, future-ready leaders have deep levels of human intelligence that is made up of mindful, critical, and strategic thinking skills, change leadership know-how, collaboration and coaching skills, and sound ethical judgment. In their simplest form, the five demands of future-ready leadership are: *presence, agility, collaboration, development*, and *discernment*.

What makes future-ready leaders *ready* is that they understand and can leverage what makes them fundamentally human to achieve results in the age of digitization, uncertainty, and emergent change. Future-ready leaders and teams realize that their human capacities will allow them to overcome the perils and maximize the possibilities of 4IR changes at work and in life.

The tools, frameworks, and resources in this book will help individual leaders and their teams deepen their understanding of what it takes to meet the demands of leading in smart, connected organizations and communities. These resources are designed to be practical and to empower leaders to take action and begin the process of future-proofing their organizations, communities, institutions, and/or workplaces.

In addition, the resources in this book are intended to help educators, coaches, and human resources (HR) professionals start developing the next generation of leaders around the five demands of future-ready leadership. AI, robotics, and the workplace of the future are new to everyone. Therefore, whether you're 25 or 75 years of age, this book offers insights and advice to help you make sense of and decide how your organization *should* approach fourth industrial revolution challenges. The tools and recommended actions provided in each chapter are designed to spark dialogue, discussion, and debate among the people in your community or organization, resulting in action that generates prosperous shared futures.

Creating a future-ready organization is a strategic process with a set of capabilities that help leaders reduce the risks and maximize the potential of 4IR mega trends such as AI, 3-D printing, and digitization. The future-proofing process requires creativity and conversation, but it also requires discipline and structure. Leaders who are committed to building a future-ready organization are concerned with long-term sustainability. Future-focused leaders are well aware that the 4IR changes taking place today can render any organization or institution obsolete seemingly overnight. Therefore, developing your organization's future-proofing capabilities is time well spent.

Clients often ask me, what do you think *will happen* in the future of work? And, although I have opinions about how automation and robotics will play out, many of which are shared in this book, I think the larger question that leaders should be discussing is what *should happen* in the future of work? Leaders have choice and agency in creating a shared future with their followers and the communities of interest they serve. But that work must begin now.

Leadership, by definition, is about influencing, motivating, mobilizing, and coordinating people and resources to achieve a shared goal.[1] Goals are ambitions, dreams, and/or results that have *not yet* been achieved. As such, leaders—regardless of whether they lead a for-profit, nonprofit, or government organization—do so with the expectation that they are always looking ahead to the future. If you cannot anticipate future consequences and set direction for others (i.e., followers), then you shouldn't aspire to lead. Moreover, if history teaches us anything, it's that leadership and focusing on the future go hand in hand. The best leaders always have an eye on what's possible, what could be, and what *should* be.

Across the first three major economic eras, the most successful businesses and institutions have been led by people who have been focused on the future, and who do take the well-being of their followers seriously. Future-focused leaders are trend-spotters—they are able to see which new technologies are on the horizon, and they have the ability to think through how those technologies will impact groups of people and society at large. The best leaders I've worked with are critical thinkers. They reason and plan for the future using a structured thought process. This is the essence of the future-proofing process at the individual and organizational levels—it's about forecasting, speculating, ideating, and designing for tomorrow. I firmly believe that in today's smart, connected world, leaders must become more future-focused and must develop stronger future-proofing capabilities within themselves and their organizations. I had to write this book because I believe so firmly that global leadership, across all of our institutions and organizational types, is not future-ready.

I wrote this book for leaders who share my belief that leaders can and must do better in the future of work and life. I wrote this book for busy leaders like you. Although the research and frameworks in this book will be of interest to academics, educators, and consultants, this book is intended to help leaders like you become future-ready leaders and build future-ready organizations.

Whether you are planning to future-proof an organization that you've spent a career building, or you are a student in a college classroom, the strategies in this book will help you get ready to lead in the future of work and life.

The technological changes that define the 4IR are unlike any that leaders have had to navigate in previous economic eras. Fourth industrial revolution technologies are challenging us to rethink work and the *very nature of what it means to be human*. To me, the fact that 4IR technologies, particularly those related to artificial intelligence, are raising questions about the possibility of a jobless future, a renewed cultural renaissance, government-backed basic income, and new forms of regulation for things like self-driving cars is a *really* big deal. Successfully tackling these challenges will take a massive amount of awareness, understanding, and collaboration between public and private-sector leaders. For this reason, we need to start the conversation, and deepen the conversations where they have already begun, as to how to become more future-ready and to build shared futures together.

I also wrote this book for leaders who may not be aware of the massive technological disruptions that are going to soon transform our organizations and our lives in major ways. Leaders are often pulled "into the weeds" of the organizations that they lead because of the complexities of the problems and uncertainties that their organizations face. Sometimes it seems like everyone in an organization has their head under the hood of the car, and, as a result, no one is looking at the road map or out the windshield. Consider this your wake-up call! This book is the competitive advantage that you've been searching for to build a thriving organization amid rapid 4IR change.

What Is the Fourth Industrial Revolution?

What I've referred to as the 4IR is not something made up—it's not an academic theory. It's an economic fact that is supported by solid economic and technological evidence. Although it may sound like something out of a *Star Wars* movie, the 4IR is here and now. The digital age is both promising and perilous and, as leaders, it's your job to make the most of it and to protect your organization from it.

Just as the mechanical loom and steam-powered engine marked the first industrial revolution, the mega trends and technologies that already exist, such as the photonic glasses described in the beginning of this chapter, have marked the earliest phases of the 4IR. Because we're in the early phases of the 4IR, it's not too late to start the future-proofing process for your organization and your workforce. You can start building tomorrow's leaders right now.

According to the World Economic Forum (WEF), the 4IR is the fourth major industrial era of human history.[2] Other authorities on the future of work have referred to the 4IR as the second machine age or Industry 4.0.[3] These authors have begun raising awareness among leaders about specific 4IR technologies and their potential impact on the workforce and economic growth. However, this book is the first of its kind to focus specifically on how the 4IR workplace will impact leadership itself.

The second machine age is different than the first machine age (i.e., computers and early Internet technologies) because of a great convergence between the digital, physical, and biological worlds. This great convergence of mega trends will bring technologies, materials, and services to the market in the next three to five years that will create exciting new experiences, products, and possibilities for the human race. And the 4IR will also bring plenty of risks, disruptions, and challenges that leaders must rise up to address.

A (Very) Brief History of the 4IR

To understand what the 4IR is and why it matters to your organization, you need to consider the 4IR in its historical context.

The **first industrial revolution** (1IR) took place over a hundred-year period in Europe and North America. Key dates associated with this era are the 18th to 19th centuries. During this period, people moved from agrarian (i.e., farming) societies to urban societies aimed at production and modernization.

A key innovation associated with the 1IR occurred between 1813 and 1814 in the United Kingdom. An engineer named William Hedley built "Puffing Billy." This was the first steam-powered engine that was built to haul coal from the mine at Wylam to the docks at Northumberland. Steam power quickly spread and gave rise to iron, railroad, and textile industries, which created multiple spin-off industries and made many large fortunes.

The **second industrial revolution** (2IR) is marked roughly by the period of rapid economic growth that began in 1870 and continued until the beginning of World War I (1914). In this era, growth in new industries fed the major cities emerging around the world. Steel, oil, gas, the expansion of railroads and telegraph systems, factory production lines, and the advent of electricity grew into heavy industries that built empires. Names like Carnegie, J. P. Morgan, Charles Schwab, and Rockefeller shaped this era in the United States.

Interestingly, the 2IR lasted only four and a half decades—less than half the time of the first revolution! Through this period of rapid innovation, the 2IR ushered in new technologies like the light bulb, the phonograph, and the internal combustion engine. In my hometown of Detroit, Michigan, men like Ransom Olds and Henry Ford introduced assembly lines in their automobile factories. Ford's improvement upon Olds' factory allowed him to crank out Model T's in about 93 minutes. A new industry was born, and the lives of many families in Detroit were changed forever—including those my grandparents!

Enter the **third industrial revolution (3IR)**. This revolution began in the late 1950s and continued into the 1970s. In 1958, Jack Kilby, a newly employed engineer at Texas Instruments, convinced management that he had solved a circuit design problem called the "Tyranny of Numbers." Kilby's solution would allow Texas Instruments to manufacture all of the components of a circuit on a single piece of semiconductor material, increasing the performance of the "integrated circuit." In 1959, the patent for "miniaturized electronic circuits" was filed. Kilby is credited as the co-creator of the integrated circuit, which we now call the "microchip." In 2000, Kilby won a Nobel Prize in physics for his contribution to microchip technology and its commercial applications.

The rest, as you know, is history: personal computers, Internet, digital information, and mobile communication technologies. Names like Gates, Jobs, and Wozniak have left their mark on the face of human history. But personal computing and digital connectivity are just the beginning.

The **fourth industrial revolution** is building on, and even disrupting, 3IR technologies and business models. It represents a great convergence and advancement of 3IR technologies, resulting in a deep integration of new capabilities that are reshaping the physical, biological, and social worlds. New 4IR technologies will be integrated into every facet of society and our physical world. Sensors will provide a digital layer on top of land and water, and even in air. These sensors will be everywhere, generating vast amounts of data that can be analyzed, utilized, and monetized in creative ways. If you own an iPhone, you're producing reams of data with every step you take and every click you make.

These technologies and others like 3-D printing, artificial intelligence, the Internet of things (IoT), robotic process automation, nanotechnology, and photonics are creating breakthroughs in every industry at a pace that was unimaginable even five years ago. The challenge that the 4IR brings isn't a lack of technology; it's our human ability to cope with, collaborate around, and lead others through this revolution. As WEF chairman and world-renowned economist Klaus Schwab has said, the societal changes that accompany these technologies will require *responsive* and *responsible* leadership.

Leaders can derive some important lessons from this very brief history of 4IR.

1. Lesson #1: There is overlap between each of the four industrial eras in human history. Industrial revolutions don't have clear beginnings or ends. These periods of history and their social impact blur and blend. Given this, one could argue that the global economy is still immersed in the third industrial revolution. Such an argument, however, is shortsighted. Future-focused leaders who are looking at the big picture realize that the 4IR is here, and they're taking action to adapt and to respond to the new leadership requirements of the digital era.

2. Lesson #2: Each of these industrial eras brought with it significant advancements that have fundamentally transformed society. Technologies of the

previous industrial eras impact where people live, how they move about, what they do with their time, their quality of life, their well-being, and connections among communities in which they live and work. Each industrial revolution introduced new opportunities and new challenges that leaders had to learn to respond to. The true leaders of each industrial era have been courageous pioneers who are focused on building a better future.

3. Lesson #3: The speed and velocity of each industrial revolution is increasing. In terms of speed, the first revolution lasted 100 years, the second lasted 45 years, and the third arguably less than 30 years. In terms of velocity, the 4IR is unique because it's taking global economies in new directions, creating new business models, and fundamentally disrupting the business models of 2IR and 3IR industries. These directions are more complex, more automated, more transparent, and raising more vexing ethical questions for leaders than ever before.

4. Lesson #4: This brief history teaches us that leadership plays a significant role in shaping industry and society's response to the changes introduced by new technologies. People have always been—and will always be—at the heart of each industrial era. These are people with ideas and vision. Industrial leaders have been able to mobilize resources, build their workforce, and sustain businesses and institutions over time. Most have been people who are trying to solve problems and make life better as evidenced by their philanthropy. As 4IR technological mega trends spread, leaders must not lose sight of the fact that humans are, and must remain, at the center of 4IR social change.

What Does the 4IR Mean for Leaders, Followers, and Organizations?

Leaders are already starting to discuss the new leadership demands that the 4IR will create for business, government, and society. In late 2017, the WEF brought together a forum of thought leaders from science, industry, and government. The WEF meeting took place in Davos, Switzerland, and the topic of discussion was the 4IR.

Leaders at the WEF meeting in Davos characterize the 4IR as a time of great promise and potential perils. The promises of the 4IR include an explosion of "new-collar" jobs and network-centric business models like Uber, Airbnb, and Instacart. And, yes, with all of the promises of the 4IR come new challenges, risks, and growing threats to job security, job training, innovation, and the human experience of work.

The challenges of the 4IR are no less daunting than the promises. They include a growing disparity between social classes, issues of data and national security, and biomedical quandaries, such as who gets to control the limits of human genetic engineering? Who has access to basic health care versus aesthetic health care? And how will society carry on if automation continues to eliminate minimum-wage—and middle-class—jobs?

The gap between people with access to good jobs and new technologies may well accelerate as automation and artificial intelligence make many blue-collar and white-collar jobs obsolete. Increasing unemployment or unequal employment could feed growing public distrust of institutions, government, global corporations, and the leaders who govern them. And even if these drastic employment challenges are overcome, leaders will still have to shift their digital strategies and their people strategies to create collaborative advantage over their competition and respond to changing customer demands.

Consequently, leaders across public, private, and social sectors must learn to create collaborative advantage through new means of attracting, developing, redeveloping, and retaining their employees. Teams will need leaders who can facilitate innovation and collaboration at faster rates, and with greater impact, as strategies and structures through which work in the digital age change. Decision making will have to become more local, workforces more agile, and leadership more human-centered and consultative than ever before.

On an individual level, future-ready leaders and followers must become smarter, faster, and more connected. Similarly, organizations must become more agile and optimize learning from failure and big data analytics. And if leaders want access to employees' best ideas, talents, and skills—assuming that organizations will still need humans, just smarter ones—then leaders must get very good, very quickly, at meeting the wants and needs of a highly connected workforce. How do we know that leaders will need to adapt to these changing demands? Because it's already taking place.

Smart, Connected Leadership: A Strategy for Collaborative Advantage

This brave new world, defined by 4IR mega trends, requires a brave new leadership mindset and approach. Although no one can predict the future with absolute certainty, the early macro-economic and behavioral economic indicators suggest the following:

- The people side of organizations will be impacted by 4IR technologies in a major way.
- Jobs will change and will involve more automation and human–machine collaboration.
- The pace of change will increase rather than decrease.
- Demand for talents fit for new jobs will outstrip the supply of that talent.
- Collaboration, consolidation, and convergence will become strategic differentiators.

- New collaborative models of shared power will emerge in organizations.
- The culture of organizations will change as automation, mobility, and speed of decision making increase.
- The need for employee education across the life span will increase.
- Leaders will be faced with huge ethical decisions in business, government, education, and life.
- Everyone will demand more transparency (i.e., customers, employees, constituents, peers, etc.).
- The workforce of the future will be a smart, connected workforce of creative analytical types.

These certainties are by no means exhaustive, but they paint a pretty compelling picture of how leadership itself will have to change and evolve. Leadership in the next 100 years will require a better quality of connection between people. Leaders will have to spend more time investing in their relationships with peers and followers. They'll have to be more present with their followers. They'll have to communicate more frequently and more authentically than ever before. How leaders "show up" to inform, inspire, and set goals will be more important to their success than ever before.

Simply put, future-ready organizations will demand a different set of behaviors and a different mindset from leaders. The new demands of 4IR leadership include:

- managing complexity with greater presence of mind, allowing leaders to engage in complex thought and action;
- collaborating and rethinking power and influence in their relationships with peers, followers, and networks;
- accelerating organizational change while ensuring employee resilience and well-being;
- learning continuously and developing other people faster; and
- making purpose-driven decisions with greater efficiency, effectiveness, and ethics.

I call this new model of influence and inspiration, **smart, connected leadership.** It is a mindset or way of thinking about the relationship between leaders and followers in the future of work. Smart, connected leadership requires learning new skills and behaviors that are fit for leading in the digital age. Developing a smart, connected mindset means letting go of conventional wisdom, old mental models, and even stereotypes that people have about leadership, which have been around for centuries. Old ways of thinking and behaving as a leader (i.e., with a top-down mindset) will no longer serve you, your team, or your organization in the future of work. It's time for a change.

You can be certain of three things about leading in the digital age:

- Historically, the "leadership development industry" has failed at preparing leaders. As Harvard University's Barbara Kellerman has noted, despite leadership having been studied for more than 100 years, little evidence proves that leadership development professionals know how to effectively teach and develop better leaders.[4] Training and coaching often fail to produce desired effects, and they aren't necessarily the *best* answers for preparing tomorrow's leaders for 4IR challenges. Promising alternatives to traditional leadership training include things like leader-to-leader dialogue, targeted experiential learning journeys, and customizable learning platforms that are tailored to a person's passions, talents, and purpose. The tools in this book will help foster that leader-to-leader dialogue. And the online communities that support this work provide a space for deeper conversation and learning about building future-ready organizations.

- The mega trends that define the 4IR are moving faster than expected, and organizations that embrace a brave new leadership strategy will survive and reap the benefits of the 4IR. The model presented in this book will provide you with a road map for developing tomorrow's leaders based on the best of what is known about leadership and human development.

- Finally, based on the data available, you can be certain that our current models of leadership and governance are inadequate for answering many of the questions that the 4IR raises. And these are no small questions; they're massive questions like, what regulatory body should oversee CRISPR, a gene-editing tool for correcting genetic mutations? And what data should and should not be shared through blockchain technologies? Given these questions, this book will help promote dialogue among leaders who influence work and life across diverse segments of society.

This book provides an innovative and bold vision for meeting the global leadership needs at this critical nexus in human history. Without a doubt, a brave new set of leadership strategies, mindsets, and behaviors are needed. This is the most important conversation that leaders around the world should be having. The future is now, and it demands a leadership response.

The rest of this book is designed:

1. to help you understand the 4IR's impact on leadership itself;
2. to create space for dialogue with other leaders and the teams with whom you work; and
3. to equip you with strategies and tools for future-proofing your organization and your career as a leader.

Toward a Smart, Connected Model of Leadership

In the following chapters, I will provide you with a new *leadership model*, development strategies, and practical tools designed to help you lead others through fourth industrial revolution challenges. This new leadership model is future-focused and grounded in cutting-edge research and observable technological and social mega trends. What this means is that the smart, connected model has been developed using the best science and data available today to support you in building a future-ready organization.

Today's data allow us to predict tomorrow's leadership challenges. As the case studies covered in each chapter illustrate, 4IR leadership challenges already exist, but they are not insurmountable. They do, however, require leaders like you to view the science and art of leading with a different mindset. To do this, you will need a new leadership framework and set of strategies to put into action.

To help you develop a new leadership strategy and practice, I will use a simple structure for each of the remaining five chapters in this book. First, I will introduce a client case study to define the theme of each chapter. Next, I present concepts and data to deepen your knowledge base and understanding of each new 4IR leadership demand.

In the second half of each chapter, I outline a set of tools and techniques to help you deepen your repertoire of leadership strategies. Each chapter provides guidance on how to use these tools and techniques to become a smarter, more connected leader. These tools, frameworks, and powerful questions are intended to foster self-reflection, leader-to-leader dialogue, and, most importantly, to help you take action to start future-proofing your organization or institution.

Finally, each chapter includes a chapter summary with recommended actions and tips for managing your accountability. In total, this book contains more than 25 development tools and 20 recommendations for action that you can put into practice today. If I were to teach your leaders how to use all of the tools in this book in person, it would take a week just to learn them. As we all know, time is money. So, I believe that there is tremendous value in the book that you're holding in your hands. These tools have taken me nearly 20 years to develop and refine, with more than 50,000 hours of research, practical experience, and advising thousands of clients on becoming better leaders and leading change in their organizations.

In each section of this book, I will help you "aim" these development tools at one of the requirements of future-ready leadership. These include:

- Presence
- Agility
- Collaboration

- Development
- Discernment

Let the Journey Begin

The global economy and the very fabric of our society are changing at a tremendous pace. The nature and velocity of these changes require a new leadership mindset and new leadership behaviors. My purpose is to help leaders like you develop a new leadership approach for overcoming the challenges of leading in the digital age. This approach consists of:

1.   a new **framework** that is responsive to 4IR leadership demands,
2.   a new **mindset** about the relationship between leaders and followers, and
3.   new **strategies** that can be implemented today.

According to Klaus Schwab, WEF executive chairman, "the new technology age, if shaped in a responsive and responsible way, could catalyze a new cultural renaissance that will enable us to feel part of something much larger than ourselves—a true global civilization."[5]

The smart, connected leadership framework and strategies that follow will improve your ability to shape 4IR in the very *responsive* and *responsible* ways that the issues of our day demand. Let the journey begin!

CHAPTER TWO

Presence

Gerry is a senior vice president for a large publicly traded company. Today she's jumping out of bed at 5:00 a.m. after attending an 11:00 p.m. conference call last night. She grabs a cup of coffee before going to see her personal trainer. While in line at Starbucks, she deletes a few emails, checks out the headlines on her *Wall Street Journal* app, and "likes" a friend's vacation photo on Facebook. The hour that she sets aside each morning with her personal trainer is one of the only outlets that she has for the tremendous amount of pressure she is under at work.

As Gerry makes her way into the office, she responds to more emails at stoplights. She arrives at the boardroom a few minutes before her 7:30 a.m. meeting with the executive committee. The executive committee is working through the details of an acquisition of a rapidly growing start-up whose capabilities will skyrocket Gerry's business unit's market share and competitive advantage. As the meeting begins, most of Gerry's peers are attentive, but are distracted and furiously typing on their tablets by 7:45 a.m. The acquisition is nearly finalized and Gerry is starting to think about what this will mean, long-term, for the business that she leads.

Gerry has been working on structuring this acquisition for the last three months. To prepare for the close of this transaction, she has spent a great amount of time away from her home and family. She missed most of her daughter's senior year soccer season and hasn't had much time for her relationships with her husband or their friends. When Gerry shared how she was feeling about her "busyness" with a close friend and colleague at work, he said, "Congratulations, if you're struggling to manage your home life, that means you've finally made it in our company culture!"

During the executive meeting, while trying to focus on the financial structure of this $250 million acquisition, Gerry cannot stop thinking about the fact that the more she succeeds at work, the more it feels like she's failing in life. As Gerry ponders her work–life quandary, she's pulled back into the room when her CEO calls on her. Gerry provides her colleagues with a five-minute brief on some financial analyses she's been working through.

When Gerry finishes, she answers a few questions from her team, and she glances over to her colleague Stewart, the CIO of her company, who is furiously typing away on email. Stewart is caught off guard when Gerry turns the floor over to him for his post-transaction IT integration briefing. Gerry sits down as Stewart bumbles through his introduction on the status of post-acquisition integration plans. As Stewart talks, Gerry picks up her iPhone and reads an email from her daughter's soccer coach; she's won an award for Best Defensive Player of the Year. It's now 8:10 a.m. and Gerry has another important meeting at 8:30 a.m.

"The decision that we really need to make," Stewart asserts, "is whether or not to roll the new company onto our system right away or let them run with their comparable system for the first year. This would help ease some of the pressure on their people during the first year. I mean, we're going to probably be laying off 20 to 30 percent of their IT staff anyhow. Delaying would give them a little bit of a break."

Members of the committee around the room nod, but clearly need more information. "Ideally, before we leave today, it'd be great if we could make that call so that I can have my team finalize our time lines and staffing for the next 6 to 12 months." Some members of the committee express arguments on both sides of this issue. It's now 8:25 a.m. Members of the committee seem indifferent, and some are still distracted by their emails. *Defensive play of the year, I'm so proud of her,* Gerry thinks to herself.

One member of the team stands up and says, "I'll vote with you either way on this, Stew. Sorry, but I have to run to my 8:30." Stewart calls for a quick show of hands. The committee agrees to delay the system integration for the first six months. Gerry starts gathering her notes for her next meeting with one of her biggest customers. Meeting adjourned.

This vignette, based on my observations with clients like Gerry, is intended to highlight the frantic pace of life of many 4IR leaders. The many decisions that they face are getting more complex with higher stakes. It also illustrates the intense pressure that leaders of all types are under, and how such pressures create pushes and pulls for attention, consideration, and deep thought. The world of work has never been defined by such complexity and speed.

These pressures are common among all types of leaders (e.g., executives, middle managers, and front-line supervisors). And these types of complex decisions apply to both for-profit and not-for-profit leaders. If leaders aren't dealing with multimillion-dollar pressures from boards and shareholders, then they're dealing with pressures to raise funds for their nonprofit or increase student enrollment for their university. Pressure is pressure, and it's all relative. What differentiates high- and low-performing leaders is how they respond to these pressures.

If you're like many of my clients, working 12-hour days is probably common for you. You are probably getting between 4 to 6 hours of sleep, less than what is recommended for optimal health and brain functioning. I'm sure you're constantly connected to your cell phone, and your day is probably filled with distractions and disruptions. This onslaught of constant connectivity and fast-paced activity can feel like utter chaos at times. Many leaders tell me that their workweek feels like chaos most of the time. If "chaos" and a feverish pace of "busyness" describe your work and life, then you are not alone.

According to the World Health Organization, this stress-inducing work pace costs American businesses an estimated $300 billion per year. Most of this cost comes from lost productivity, absenteeism, and sickness. Stress is literally killing leaders in every industry. One Fortune 500 executive I interviewed for this book described the pressure that leaders in his organization face like this: "If we don't figure this thing out, we're just going to have a bunch of dead executives in their 40s." Indeed, over the past 30 years, well-being researchers have observed an 18 to 23 percent annual increase of self-reported stress among men and women, respectively.

My goal in this chapter is not to criticize the pace or highly distracted lifestyles that many people lead in their work and home lives. My goal, instead, is to argue that this highly distracted and stress-induced way of working is bad for business. In other words, your "busyness" and lack of what I call "leadership presence" needs to change in the new world of 4IR work. Effective leadership and elite performance in the digital age presents a new set of complex cognitive demands that will require leaders to change how they work. Specifically, fourth industrial complexity requires leaders to be more mindful about both the content of their work (i.e., what they focus on) and their work flow process (i.e., how they get work done).

Smart, connected leadership is defined by being fully present with the work at hand. In this chapter, I unpack the idea of *presence* and provide some practical strategies for improving your focus, presence, and resulting performance. These strategies are not "personal productivity" tips about calendar management or sorting through your cluttered email inbox. Those are tactics. What this chapter provides you with is a new approach for focusing on what matters most. This approach starts with your own habits of thinking, feeling, and leadership behavior. By improving your leadership presence, you can start to create a smart, connected team where deep work and true performance are valued more than mere activity or "busyness."

Future-Ready Leadership Demands Presence

Presence is a term that has long been associated with leadership. I'm sure that you've heard people say things like, "He just needs to work on his leadership presence," or "She just lacks the executive presence that this role demands." This type of presence refers to the "poise," "impression," or "personae" a leader projects. This is not the type of "leadership presence" that I'm tackling in this section.

The type of leadership presence that I advocate for in this chapter is about presence of thinking, feeling, and acting. *Presence* is a term that is often associated with mindfulness, meditation, and various Eastern philosophies. But I don't want to get too academic about what mindfulness is. I want to keep the definition of presence simple and practical.

Simply put, presence refers to the state of existing in the moment, being aware and attentive, and free from the noise and "clutter" in your head. Presence is the opposite of chaos. Chaos is disorderly, uncertain, and messy. In a world of work that often feels chaotic, presence is more important than ever because even when things are chaotic, presence of mind may be the only thing that a leader has control over. Presence is ultimate awareness, focus, and control of your thoughts, feelings, and behaviors.

When people work with greater presence, whether they are frontline employees or an overscheduled executive like Gerry, they bring more of their cognitive, creative, and experiential wisdom to the issues and decisions at hand. Presence is a state of mind and a state of being that enables more productive responses to challenges at work and in life. It's a state of mind that can give rise to our best self and to that which makes us fundamentally human. Presence can be a most valuable capacity to develop within yourself and others to make yourself, and your role in your organization, future-proof.

In their 2011 book, *Race Against the Machine,* MIT economists Erik Brynjolfsson and Andrew McAfee discuss how the fourth industrial revolution's knowledge-based economy is creating a "Great Restructuring" of jobs and work.[1] Automation is a major driving force behind the Great Restructuring. As economic data since 2011 have shown, repetitive jobs that involve handling "things" or repetitive manipulations of data can be easily outsourced to machines.

As a result, the Great Restructuring is impacting when, where, how, and by whom (or what machine) work is getting done. This means that the most successful leaders and individual contributors in the next phase of the fourth industrial revolution will have a certain set of nonautomatable skills. You may, as Northeastern University's President Joseph Aoun does, call these skills "robot-proof."

Aoun has written a great book titled, *Robot-Proof: Higher Education in the Age of Artificial Intelligence.*[2] In this book, Aoun explains four new cognitive functions that leaders must master in the future of work: critical thinking, systems thinking, cultural agility, and entrepreneurship. According to Aoun, the demand for these cognitive functions will increase in in the age of artificial intelligence (AI). At the same time, tasks that can be easily automated will decrease in demand because it will be more sustainable for businesses to outsource these tasks or have robots (or AI) do this work. In other words, everyone in the workforce should be thinking about how they can create a robot-proof career.

The jobs that are most susceptible to automation are those that are repetitive and involve touching, handling, or manipulating "things" or basic information. For example, moving boxes in a large warehouse can now be easily automated. Driving a truck full of goods, someday in the near future, is likely to become automated. Similarly, producing financial reports, diagnosing MRI data, and even managing financial portfolios are all susceptible to automation. These are economic facts that are simply driven by the law of supply and demand. I'm not a "rise of the robots" doomsayer; I'm just realistic. Also, my work with Fortune 500 clients has provided me with an up-close view of the types of automation pilot studies that many organizations—big and small— are conducting. The net impact of the automation pilot studies that I've been privy to all seem to have the same outcome: a fundamental redesign and restructuring of work.

The Great Restructuring of work is simply making certain skills more valuable than others. Essentially, demand for easily automated work (i.e., "mindless work") is decreasing, and demand for "mindful" smart, connected work is increasing. The result is that making a decent living and thriving in a 4IR economy is going to require smarter, more connected work, which means employees are going to have to be taught how to be more present to do deeper, more mindful, work well.

You might say that the Great Restructuring is resulting in a great "resorting" of skills and talents. The people who will be most successful in the new world of work will be those whose skills are not easily automated. These include creative skills, complex problem-solving skills, communication and collaboration skills, and an ability to learn hard things fast. The new world of work is going to increase the value of those who are fast learners, and those who can engage in deep thinking and problem solving.

Leaders who understand the power of presence, and its impact on peoples' ability to think, feel, and behave in certain ways, will be able to leverage their employees' presence to create, to innovate, to solve problems, to learn faster, and to change faster than their competition. Similarly, employees who learn how to harness their presence of thinking, feeling, and behaving can take greater control of their careers, and can improve their physical, financial, and social well-being in the process.

Presence, however, is something that employees and leaders alike will have to learn in the digital age. As the case at the beginning of this chapter shows, leaders like Gerry—if we assume she is in her mid-50s—have spent their entire careers in chaotic organizations. Most companies have an outdated notion of what productivity and value are.

Productivity is often defined in terms of units produced in a given period of time (e.g., widgets per hour). It seems our modern-day widgets are emails and meetings attended. These notions of productivity and efficiency stem back to work by Frederick Taylor, who was an American engineer and one of the founding fathers of scientific management. In 2001, honoring Taylor's influence on modern management theory and practice, Fellows of the Academy of Management voted Taylor's 1911 book, *Principles of Scientific Management*, the most influential book of the 20th century.

Indeed, Taylor's thinking has had a tremendous impact on modern management. He is often credited as pioneering the entire field of industrial engineering. His methods of standardized work, "best implements," supervision, scientific selection and training of employees, and so on, lives on in the knowledge economy and in 4IR organizations. Taylor's theories and principles have shaped the global mindset of how modern organizations are run—and they've probably even shaped how you think about leadership and management.

But here's something you might not know about how Taylor viewed work. He didn't put much stock in the ability of most employees within an organization to engage in mindful, smart, connected work. Taylor believed strongly in division of labor, especially the division between management and workers. One of Taylor's four principles of work was that work should be equally divided between managers and workers so that managers could scientifically plan the work to be done (strategic work) and workers could perform the tasks (tactical work). In other words, managers and leaders know best.

According to Taylor,

Now one of the very first requirements for a man who is fit to handle pig iron as a regular occupation is that he shall be so stupid and so phlegmatic that he more nearly resembles in his mental make-up the ox than any other type. The man who is mentally alert and intelligent is for this very reason entirely unsuited to what would, for him, be the grinding monotony of work of this character. (p. 59)[3]

Although this quotation doesn't paint a very favorable picture of Mr. Taylor, it does illustrate how Taylor conceived the difference between management and labor. Managers were to be "mentally alert" and "intelligent." Workers were thought to be "stupid" and "phlegmatic."

Modern leaders might not survive too long these days if they see their people as being as dumb as an ox, but the residual impact of Taylor's thinking on the separation between management and labor, strategy and tactics, and hierarchical control are alive and well in many organizations today. For better or for worse, Taylor's principles of scientific work have left an indelible mark on how we think about and structure work. Moreover, Taylor's thinking about performance and productivity have shaped what many leaders value.

Today, however, performance and productivity are often equated with how busy a person is. How packed is your calendar? How many hours of activity have you logged? How many emails do you send and receive each day? Cal Newport is Georgetown professor of computer science. In Newport's 2016 book, *Deep Work: Rules for Focused Success in a Distracted World*, he traces the prevalence of "shallow" or mindless work in the modern workplace back to Frederick Taylor. Newport defines deep work as "Professional activities performed in a state of distraction-free concentration that push your cognitive capabilities to their limit. These efforts create new value, improve your skills, and are hard to replicate" (p. 3). According to Newport, deep work is rare in modern workplaces. Modern work is filled with tweets, pings, likes, emails, and all sorts of other distractions that shift our focus from what really matters.

Paradoxically, despite the prevalence of shallow work in 4IR workplaces, Newport argues that deep work is more valuable for individuals and organizations. Here's why: deep work—regardless of your business or field—by definition creates new value and enhances your skills. Therefore, on an individual level, if you are a musician, your deliberate practice of your instrument helps you sharpen your skills and produce more creative works.

Similarly, if you're an academic like Cal Newport, you will publish a slew of academic articles, write four books, maintain a popular blog, and still shut your computer down around 6:00 p.m. As a result, Newport creates a virtuous cycle for himself—creating more widely read work, getting new ideas, and sharpening his craft as a writer and productivity and performance thought leader. But, Cal Newport's deep work doesn't just benefit him; it also benefits his organization. I mentioned Newport was a faculty member at Georgetown. The more impact in the world that Cal Newport creates, the better off Georgetown will be. By having faculty members like Newport around, they will attract more students, generate revenue from enrollment, and hopefully attract the attention of big donors. Who knows, perhaps Cal Newport will one day create an endowed professorship or scholarship for students with some of his book royalties!

As with anything that is rare, deep work is more valuable. By being fully present with our work, and engaging in focused mindful work, knowledge workers can benefit themselves and their organizations. So why is deep work so rare? According to Newport one of the reasons why deep work is rare is because "busyness" has become a proxy for productivity. According to Newport, "in the absence of clear indicators of what it means to be productive and valuable in their jobs, many knowledge workers turn back toward an industrial indicator of productivity: doing lots of stuff in a visible manner" (p. 64).

Thinking back to Frederick Taylor's scientific management and Cal Newport's explanation of busyness as a proxy for performance and productivity, you might ask yourself, How far have our mental models evolved since 1911 when Taylor wrote *Principles of Scientific Management*? Clearly, now more than ever, the knowledge economy is being driven by the creative talents and skills of focused employees with good ideas and rare skills. Yet, organizations often reward and value—implicitly, if not explicitly—non-value-added, non-deep, and mindless work. How busy are you? How many hours have you logged? How long does it take you to respond to an email? Do you ever truly unplug or are you always connected?

Smart, Connected Presence of Mind

Up to this point in the chapter, I've argued that leading and working in the digital age requires a deeper kind of thinking and creative work for leaders and the people they lead. If for no other reason, presence of mind is required because the problems that organizations face are more complex, and the ethical stakes are higher than ever before. However, a number of social, culture, and organizational forces hinder and distract leaders from being mindful and present with themselves, other people, and their work. The strategies and techniques in the remainder of this chapter are intended to help leaders improve their *actual state of being* present amid all of the digital distractions around them.

Now, I admit, "state of being present" sounds a little New Agey, and it's fine if you think so too. But let me be clear, being present amid the chaos and uncertainty of the day doesn't mean that you have to meditate in full lotus while sitting atop your desk. Although if that works for you, helping you engage in more mindful, deep work, then—by all means—go for it.

Leadership presence simply means that you need to consciously create extended moments (as well as micro-moments) of being present with other people, with customers, with employees, and with your own thoughts throughout the day. This takes intentionality, ritual, and practice. This will not only improve your own individual performance achievement and sense of well-being, it will also help you create an environment in which other people can freely make the choice to develop greater presence in their own work and lives.

The workplace of the future will challenge leaders to get better at three important aspects of leadership presence:

- Presence of thinking
- Presence of feeling
- Presence of acting

Elite performance requires a sound state of mind and emotional state. Just as physical activity is good practice for maintaining a healthy body, presence of thinking and feeling are good practices for managing the stressors and complexities of life in the digital age. In consolidating my research for this chapter, I was reminded of a powerful story that a well-decorated navy captain shared with me. This story is just one of many that illustrates the important connections between presence of thinking, feeling, and acting.

The Captain's Leadership Presence

A few years ago, I was consulting with a government organization, and I met a retired navy captain. I was interviewing him on what life and professional experiences had impacted his thinking about leadership and how he leads. He told me a story about when he was a young commander in the navy, and he was navigating his ship into some potentially hostile territory. Things had not yet gotten bad, but looked like they might get "messy," as he put it.

While he was standing on the bridge of the ship, the captain said that he placed his forehead in his hand and let out a big sigh. He was stressed out about the weight of the situation in front of him. And he also shared that he was uncertain, and just plain scared, about how to handle the situation.

When the young leader looked up, he said that he saw his entire crew looking at him wide-eyed and full of fear. As he spoke, the distinguished captain recalled, "In that moment I stood up straight, gave a few orders, and learned a valuable lesson about the power of being present as a leader." The lesson was that if you don't have presence of mind, you may not be meeting the needs of your crew. Ultimately, the captain and his crew made it home safely.

But the captain's story illustrates an important point about how our emotions, thoughts, and behaviors are all tightly bound up in these moments of leadership presence or lack thereof. The gravity of the situation that the captain found himself and his crew in created a tough moment, which challenged his mental presence. While he was working through the complexity of the situation, he essentially forgot that he was "on" and visible to his crew. As the captain's mental uncertainty and emotional state escalated, his behavioral presence (i.e., putting his head in his hand) communicated precisely what he was experiencing internally—fear and uncertainty.

First of all, these responses are completely human and understandable, given the extraordinarily stressful circumstances that the captain found himself in: uncertain, complex, and dangerous. Although most organizational leaders don't find themselves under this kind of life-or-death pressure—as I stated earlier in this chapter—pressure is pressure. Great leadership presence does not require that leaders suppress or deny these innately human responses to stress and high-pressure situations. However, less extreme examples of leadership pressure, perhaps like those experienced by Gerry at the beginning of this chapter, beg the question, What's required of leaders in high-pressure situations that's different from other people? Do leaders have a responsibility to be more present for their crew? And, if so, how can they do it?

One body of research that is useful for understanding how leaders, like the captain and Gerry, can develop greater leadership presence pertains to self-awareness and self-management. Peter Salovey, the current president of Yale University, and John D. Mayer, a psychologist at the University of New Hampshire, were instrumental in advancing our understanding of how people perceive, use, understand, and manage their emotions in different circumstances. These researchers referred to this process as *emotional intelligence*.

Science journalist and psychologist Daniel Goleman made the concept of emotional intelligence popular in 1995. Three emotional intelligence concepts are useful for making sense of the captain's leadership presence. These are:

- Self-awareness
- Self-management
- Relationship management

I introduce the notion of emotional intelligence, or EQ as it's often called, for two reasons. First, the science behind EQ is sound. Second, the EQ constructs and related leadership development tools are fairly widespread in modern organizations. As such, EQ research and practical tools help explain why presence of thinking, feeling, and acting matter today, and will continue to matter in the uncertain future of work. Moreover, this section can serve as a reminder to leaders who have some grounding in EQ that, perhaps, it's time to dust off that EQ tool kit and training you took 10 years ago, particularly if you find yourself getting emotionally triggered like Gerry or the captain.

In the captain's case, emotional experience led to personal insight, growth, learning, and behavioral change. Once the captain gained self-awareness about the *thought, feeling,* and *action loop* (TFA loop) that was triggered by the stressful situation in the moment, he received important feedback from his crew and self-managed his behaviors differently. Moreover, the young commander learned an important lesson that was still with him some 30 years later. This feedback and learning drove a different set of behaviors that he found more constructive for getting his desired results as a leader.

The TFA loop is a foundational tool for becoming a more present and self-aware leader.

Tool #1: The Thought, Feeling, Action (TFA) Loop

Emotional intelligence requires self-awareness. The TFA loop provides a visual representation of how self-awareness and having presence of thinking can help leaders achieve their desired results. In the captain's case, he became more **self-aware** about how his emotions and behaviors impacted his crew. Initially, the captain's lack of *presence of thinking*, or *default thinking*, led to unconscious behavioral gestures. His simple gestures of placing his forehead in his hand and sighing unintentionally created fear and uncertainty in his followers, which became a critical turning point, expanding his self-awareness.

Second, the captain learned about the importance of *presence of feeling* and **self-management**. Leading amid uncertainty requires that leaders have enough

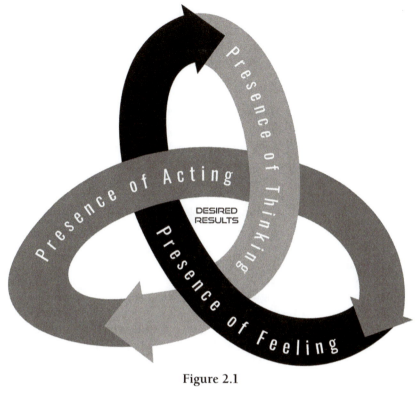

Figure 2.1

presence of mind to notice and recognize the emotions that they're experiencing in any given moment and pay attention to them. This is the essence of presence of feeling. Presence of feeling doesn't mean that you repress your emotional state. Quite the contrary, developing presence of feeling requires that you get very familiar with your present emotional state. Presence of feeling requires moving toward all the different emotions you experience when leading amid volatility, uncertainty, complexity, and ambiguity. These emotions can include, but are not limited to, fear, anxiety, frustration, anger, and even excitement, elation, and enthusiasm. Practicing paying attention to your emotions, allowing yourself to be present with them, and simply sitting with them helps you learn how to self-manage (i.e., respond) your emotions and achieve your desired results.

Finally, the captain demonstrated **relationship management** (and some quick learning). *Relationship management* is the term that Goleman applies to describe the act of using emotions to induce desired responses in others (e.g., influence, change, collaboration, teamwork, etc.). Relationship management is made possible when a leader has *presence of acting*. Presence of acting is behavioral mastery at its best, as intentional actions lead to intended results. The captain showed this in-the-moment presence when he *straightened up*, gave a few *quick orders*, and went about bringing the *crew home safely*.

The lessons learned from the captain's presence can be summarized as follows:

1. Presence of thinking leads to greater self-awareness. We need presence of thinking to get out of the default (autopilot) mode of thinking.

2. Presence of feeling supports self-management and enables presence of action. This is the type of "practical wisdom" (*phronesis*) or in-the-moment decision making that digital organizations require of leaders. Highly distracted leaders create cultures of distraction, uncertainty, and chaos. Remember, like the captain, your crew is watching you too!

3. Mastering the thinking, feeling, acting loop (TFA loop) is essential for getting your desired results. The TFA loop is a foundational concept to practice and master.

As a practical tool for managing your presence of thinking, feeling, and acting, the TFA loop can help you gain clarity about what kind of results you want/need to achieve with your peers, coworkers, and your crew. You can use this tool to practice being present with others amid uncertainty and chaos by following these steps.

First, write down the desired results that you want to achieve in the center of the loop. Your desired results might be personal in nature (e.g., to spend more time at your daughter's high school soccer matches). If they are work related, your desired results could be related to a specific business problem (e.g., improving customer satisfaction scores) or a strained relationship (e.g., improving trust with key business partners).

Once you have a desired result clearly in mind and have written it down in the center of the TFA loop, then ask yourself, or someone who knows you well, what type of presence of thinking will be required to achieve this result. Will achieving the result require that you engage in less chaotic "shallow" work, and spend more time in deep work (e.g., refining your customer engagement strategy)? Will it require a daily ritual to get you focused or keep you focused? What will be necessary to keep you grounded and focused on the one thing that you want to achieve?

As you reflect on how your thinking may be impacting your emotional presence and results, try to question the "story" that you're telling yourself about how you're showing up as a leader. This metacognitive "questioning" of your own mental models is really advanced mindset work. It may be easier for you to use a thought partner or coach to question the story that you're telling yourself about why your current presence of mind is helping you or hurting you. In either case, try to notice how old thought patterns may be getting in the way of achieving the results that you want for yourself and for your organization.

For example, if you want to improve customer satisfaction, start by asking, What are my current beliefs about customer satisfaction? What are my values around putting customers, patients, or students first? As you reflect on these questions, try to stay open to the possibility that your customer satisfaction strategy could be wrong. I know, embracing the idea that you could be wrong is radical, but give it a try! You can start by asking some "what if" questions: *What if* you screwed up by not putting a customer engagement strategy in place? *What if* it isn't your frontline employees' or managers' fault? *What if* customer preferences have changed and your frontline employees—those closest to the customer—know the solution? *What if* your middle managers or frontline employees aren't telling you how to fix the problem because they think you're too busy, or worse, because they think that you're not listening? *What if* the Taylorism mindset of "leader knows best" that you're still carrying around in your head doesn't work in the fourth industrial revolution?

Finally, by wrestling with these "what if" questions, either on your own or with a professional coach, you'll likely feel a shift in your emotional presence. This shift could be the "duh" moment that we've all experienced when we've realized that we were doing something dumb. You might also experience fear or frustration upon realizing that you have to change your thinking or behavior. Regardless of what the emotions are that "bubble" up during this exercise, sit with them, try not to judge them as good or bad. Just be present with them and notice what happens.

Staying present with emotions is a funny thing. Most of the self-help gurus say things like, "Just try to notice where the emotions are in your body." Again, it's New Agey, but admittedly, I've followed such advice and spent many hours "noticing" where my emotions are in my body. I've been amazed at what happens when I simply "watch" my emotions and stay fully present with them. I know that my experience of anger and frustration "sit" in my stomach or on my shoulders. I find that there's power in knowing this, insofar as this self-awareness leads to better self-management and better overall leadership presence.

What's more, once I've worked through the thoughts and feelings necessary for achieving desired results, knowing what to do next (action) just sort of emerges. I've found this to be true in working the TFA loop myself, as well as in working the TFA loop during thousands of hours of coaching hundreds of different leadership clients. Once leaders really decide what's most important to them and reflect on how their thinking and feeling contributes to their success or failure, they seem to be able to quickly develop a plan to guide their behavior. If the plan doesn't emerge, and you've made it this far, just ask yourself, What specific behaviors will be required of me to achieve my desired result? What support will I need? This is high-quality self-coaching for developing greater presence of thinking, feeling, behavior, and results!

Should We Believe the Captain's Story?

Although my analysis of how the navy captain quickly mastered his presence of thinking, feeling, and acting has face validity, some readers will want to see the data. What science supports my case for becoming a more mindful, smart, connected leader?

The science around "mindful leadership" has simply exploded over the last decade. Whereas 30 to 40 years ago, the term *mindfulness* called forth images of Buddhist monks engaged in transcendental meditation or hippies at a Grateful Dead show, today mindfulness is regarded as an enormously effective tool for improving leaders' health and performance.

At last count, more than 380 studies in 160 scientific journals address the psychological and physical effects of mindfulness and transcendental meditation (TM) techniques. The benefits of TM range from improved cortisol levels and reduced stress, to higher-quality relationships, engagement, and life satisfaction. Leaders can learn a great deal from TM research that will help them develop greater presence of thinking, feeling, and acting. Again, I'm not saying that future-ready leaders must practice the kind of meditation that popular media proliferate. I'm simply suggesting that developing some form of intentional practice (e.g., running, walking, getting up really early, prayer, conscious breathing, tai chi, or yoga) is helpful for keeping leaders grounded, physically healthy, and in the moment when they are faced with cognitively complex tasks. Any activity that helps to keep you connected with your greater purpose and the present moment will support your growth and performance as a leader of a future-ready organization.

How to Cultivate Presence in Self

Cultivating presence in yourself begins with identifying your purpose. If you are unclear about your purpose, then you will struggle to be fully present in your work and in your life in general. Leaders who are unclear about their ultimate "why" (i.e., why they exist, why they do what they do, why they are on the path they are on, etc.) tend to feel, and outwardly communicate to their followers, a sense of restlessness. This restlessness is particularly common among early-career leaders. Without a "why," you will struggle with identifying your desired results and maintaining presence of thinking, feeling, and behavior.

One hypothesis I have about the importance of purpose, based on in-depth coaching with hundreds of leaders and their teams, is that purpose and meaning at work help to create a *sense of coherence* in life. This hypothesis is based on the basic principles of logotherapy. The basics of logotherapy come from the work of physician, therapist, and Holocaust survivor Dr. Viktor Frankl. In Frankl's 1946 book, *Man's Search for Meaning*, he describes the basic premise of logotherapy, and how to pursue a purposeful life.[4]

Frankl's theory of logotherapy is informed by his lived experiences, specifically, the suffering he observed and endured during three years of internment in Nazi concentration camps. However, following his release in 1945, Frankl continued to deepen his work around logotherapy, holding professorships at the University of Vienna, Harvard University, Southern Methodist, Duquesne, and Pittsburgh. Frankl died in 1997.

However, Frankl's teachings are still alive and well. Logotherapy teaches us that all life is meaningful, even under the most miserable circumstances. Humans have a "will to mean" and pursue their purpose. The pursuit of purpose and meaning can never be taken away from a person. However, according to Frankl, when people lose sight of their ability to experience meaning in the things they do, the people they encounter, or work they choose, this can lead to existential crisis and neuroses of all different types.

According to Frankl, meaninglessness in life is the cause of many existential neuroses in human beings. He is said to have coined the term *Sunday neurosis* to describe the anxiety that people experience at the beginning of a new workweek. This feeling can cause people to feel depressed, hopeless, and directionless. As a reminder, these "people" aren't just unique samples from Frankl's research. They are human beings like Gerry and the captain. The people who experience Sunday neuroses are the employees and staff in your organization. They're your peers or friends and family of your peers. When people in a community lack purpose, the entire community is impacted. Thus, for Frankl and his contemporaries in fields like positive psychology, the key to building and leading high-functioning communities, organizations, teams, and individuals is purpose and meaning in the deep work we commit ourselves to.

However, when leaders fail to connect disruptive change with their organization's purpose, a crisis in clarity, meaning, and commitment can occur. Without a clear and compelling purpose, the tasks of day-to-day work can become overwhelming, and people can lose sight of what to focus on and why. Take Gerry, for example, at the beginning of this chapter. The tension between her ultimate purpose as a corporate executive and as a mother seemed to be wearing on her. To help Gerry create a sense of coherence in her work and life, and to be more fully present when she is at work and with her family, I would recommend she map her ultimate purpose relative to her goals and tasks.

Figure 2.2 illustrates the purpose, goals, and tasks hierarchy, providing clarity about how purpose can provide direction in our leadership and our lives.

Figure 2.2

Tool #2: The Purpose, Goals, Tasks Hierarchy

Keeping our purpose top of mind is critical in a world of "shallow" work and lots of tasks (e.g., meetings to attend, emails to answer, and tasks to tick off a list). When we establish a clear "why" to guide our goals and tasks, the "what" and "how" of managing an otherwise chaotic array of priorities become much more manageable. Purpose serves as an anchor to which our daily tasks should align. Continuously seeking presence of thinking and feeling through daily practices keeps us connected to our purpose and enables us to "sort through the clutter" or "tune out the noise" in rapidly changing organizations.

Having presence of mind about one's purpose can also be very useful to leaders in creating purpose-driven teams and organizations. Future-ready teams and organizations are connected by a shared purpose and shared meaning around why they do what they do. This helps leaders and teams to be smarter about designing their work and prioritizing tasks. Without a clear and compelling purpose, it's hard for teams to be mindful and present with the work that matters most.

Creating a team or culture of presence starts with leadership. Leaders must have clarity of purpose and presence of mind before they can help others develop purpose and presence. Just as flight attendants say during the pre-flight safety presentation, "In the event of a loss of cabin pressure . . . be sure to *secure your own mask first* before assisting others." In the realm of presence and purpose, leaders cannot be much help to others or to the organization if they haven't found presence and purpose for themselves first.

One of the simplest and most effective techniques for cultivating presence in yourself is to use the "5 Whats." Most leaders are familiar with the *5 Whys* concept because it is a common Six Sigma and process improvement tool for determining the root cause of a problem. In my own coaching and consulting practice, I often use the 5 Whats to get at the root of an individual leader's or team's purpose. I learned this technique from one of my mentor coaches, Micki McMillan, who is an assessor for the International Coach Federation and founding partner and CEO of Blue Mesa Coaching.

For illustration purposes, let's apply the 5 Whats to Gerry's case from the beginning of the chapter. Gerry's actions (e.g., working long hours and missing her daughter's soccer matches), I suspect, are generating anxiety, and that anxiety is impacting her presence of feeling and thinking (see TFA loop).

Here's how I might use the 5 Whats to coach Gerry to greater clarity of purpose, priority, and, ultimately, presence.

Gerry: I'm just really feeling off about missing my daughter's soccer games this year. I mean, she's the defensive player of the year, and I haven't seen a single game! What a terrible mother!

Me: That's a pretty powerful assessment of yourself.

Gerry: Well what do you expect?

Me: I'd expect you to prioritize the things in your work and life that are most important to you. Let me ask you this, **what's** so great about work that you'd prioritize it over your daughter's games?

Gerry: I have a responsibility with this big acquisition.

Me: Okay, **what's** important about this acquisition to you?

Gerry: Huh. Good question. I suppose it means growth and opportunities for the people I lead.

Me: **What's** *even more* important to you than creating opportunities for your team?

Gerry: Wow [pauses]. I've thought about this once before when my husband and I decided to have children. What's more important to me than creating opportunities for my team is proving to myself that, as a female executive, I can do the extraordinary work that my mother never had a chance to do. But, I know that she could have been an amazing leader.

Me: That sounds very close to your heart. Can we go a little further with this, if that's okay with you?

Gerry: Sure.

Me: **What's** even more important to you than doing what your mother never
   had a chance to do?

Gerry: I want to give people, especially women, of all backgrounds opportunities for success in the world of business and technology.

Me: Would that be meaningful to you?

Gerry: Yes [pauses]. That's why I do what I do. That's why I sacrifice time
   with my family at times.

Me: If I may, Gerry, just one more time: **What** might be *even more* important than giving women of all backgrounds opportunities for success?

Gerry: [With a quiver in her voice] I want my daughter to know that she
   can become anything that she wants to be and fulfill her dreams.

And, cut. This sample coaching dialogue is one I've had before several times
with both male and female executives. And, make no mistake: the emotional
presence of these conversations is equally powerful for men and women. I do
not wish to cast a stereotype of Gerry because she is a female executive.

The example illustrates two important points about purpose and meaning: first, it shows how the 5 Whats can be used to help clients discover their
true purpose and to develop presence of thinking and feeling. As it turns out,
Gerry is not a terrible mother. In fact, Gerry's motivating purpose for putting
the acquisition over soccer is fundamentally to help her daughter feel empow-
ered and for her to have a good role model of success. Now, you might think
attending soccer is a part of good role modeling based on your mental models about mothering or a work–life blend. You're entitled to this opinion.

However, as this case shows, Gerry's driving purpose is to show her
daughter and women of all backgrounds that they can succeed in business
and technology. It's not my job (or yours) to judge or evaluate her purpose.
My job was to help Gerry clarify what is most meaningful to her so that she
can create greater coherence and presence in her life.

The second point that this example illustrated was the value of the 5 Whats.
When used in conversation with a thoughtful partner, coach, or friend, the
5 Whats can help leaders discover the root of their purpose and passion. This
is a powerful tool in helping leaders "sort through the clutter" of their daily
tasks and secondary goals (see Tool #2: Purpose, Goals, Tasks Hierarchy). It
provides a structure for digging underneath surface-level thoughts and feelings about tasks and goals, to get to the underlying "why." Getting to "why" is
the first step in cultivating presence of thinking and feeling about your why. If
you don't know your purpose, you cannot be fully present with it.

In this example, I used the phrase, "what's important" or "what's *even more*
important than . . ." to move from tasks to goals to purpose in just a few turns in
our conversation. In just five "whats," I was able to help Gerry uncover a deep

primary goal (i.e., helping women achieve) and something that was even more deeply meaningful to Gerry (i.e., helping her daughter succeed). This is her ultimate purpose in work and in life—or at least a pretty darn important part of it.

With this self-awareness, Gerry and I could have gone on to explore how that purpose could be used in the future to guide her thoughts, feelings, and actions. For example, we could have explored the connection between her purpose and the choices she makes in prioritizing work and soccer matches. Had Gerry wanted to go in this direction, I might have invited her to consider how her daughter feels about her missing the matches. I might have asked Gerry if she's ever shared her purpose with her daughter, or inquired about what a fully present conversation with her daughter might look like, if they were to discuss their shared purposes. This process of reflection, undoubtedly, would lead Gerry to a set of actions and commitments for living a smarter, more connected life amid the chaos and frenetic pace of her work. I've seen it happen in clients time and time again. Once leaders have a why to live for, they can manage almost any what or how. This presence and purpose will be essential modes of operating for leaders of future-ready organizations.

Finally, this case shows that it's hard to be fully present with others if you are unclear about your purpose. Daily tasks seem to lack direction and significance. Meetings and endless email chains fail to make sense when they are seemingly misaligned with your ultimate purpose. As a result, it's very difficult to stay fully present with this work. It becomes emotionally frustrating and leads to actions that lack purpose and direction. Ultimately, when you lack purpose, it's hard to be present with yourself. And if you're not present with yourself, it's nearly impossible to be present with others.

As I write this paragraph, I'm reminded of one of my favorite quotes from Mahatma Gandhi, "You must be the change you wish to see in the world." In other words, your leadership presence begins with disciplining your own mind. You must be the presence in your organization that you seek to create. If you want a smart, connected organization, then you have to become smarter and more connected to your sense of purpose and peace of mind. Disciplining your mind is the first step in cultivating a calm, confident leadership presence that creates stability amid chaos—or, like the captain's example, creates a sense of calm even when your crew is entering turbulent waters.

In helping you create a sense of calm within the chaos of fourth industrial leadership, I want to offer you a few more tools for cultivating presence of thinking, feeling, and acting. I realize that not every leader has a coach or someone they can trust to have the type of conversation I had with Gerry in the previous example. This is sad, but true. It's lonely at the top.

Therefore, the tools in the next section are designed to be "do-it-yourself" (DIY) tools, and I find them helpful for quieting my own mind. Many of my clients over the years have found them helpful as well. One word of caution, however: if accelerating leadership development for yourself or your organization is a priority, don't do it yourself. Get professionals to help. Taking on a DIY project at home is great fun and can be a nice challenge, but when you need to complete a project of great importance efficiently and well, trust professionals who are masters of their craft and their tool kit. I always tell clients, there are *tools* and *tool masters*. Or as one of my clients said, "An idiot with a tool is still an idiot." So, just because I have a hammer and a saw does not make me a master carpenter. You can use the tools in this section to improve purpose and presence in your organization. But you have to responsibly assess the urgency with which you need to develop future-ready leadership. If the future is now in your organization, then call a good architect and carpenter.

Minding Your Presence of Mind

Leaders need methods for turning off (or at least turning down the volume on) their "fight-or-flight" response. Default brain activity and fight-or-flight reactivity can lead to poor decisions and poor health. Brain science is conclusive in showing that people make poor decisions when they are stressed and guided by default rather than mindful thinking. Not to mention, over time increased cortisol levels in the blood can cause a number of physical problems in the body. This section is designed to help leaders tame the fight-or-flight responses associated with high-stress, fast-paced change.

Let's dive into some strategies and tools that will help you become a more present leader on your own. Consider my review of these tools as a menu of options for developing habits and rituals to help you become more present for yourself and others. I cannot tell you which techniques will be the best for you; that's for you to figure out. I can, however, tell you that developing presence of thinking, feeling, and acting will become even more essential in all industries as 4IR technologies increase in complexity, velocity, and volatility.

Cultivating presence in yourself begins with self-awareness. Whether you're trying to keep your head clear so that you can solve a complex organizational problem, participate in a difficult conversation with a peer, or reduce your stress levels so that you don't have a heart attack at 45 years old, developing self-awareness is a critical first step.

Presence of thinking is the *switch* that controls our awareness of our feelings and actions. Leaders who have presence of thinking, feeling, and action have a keen ability to monitor, modulate, and maintain responsive and useful emotional and behavioral states. Tool #3 is a simple illustration to remind you about the important connection between thoughts, feelings, and actions.

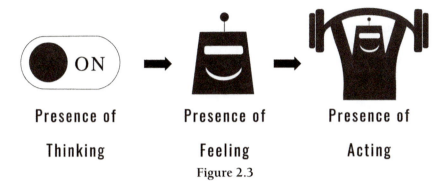

Presence of   Presence of   Presence of

Thinking   Feeling   Acting

Figure 2.3

Tool #3: Switch Your Leadership Presence on by Minding Your Mind

Your thoughts are powerful *switches* for controlling your performance and results. You must learn to master thoughts by minding your mind. Minding your mind means that you have the ability to recognize when you are losing control of your thoughts, emotions, or ability to act in ways that will get you the results that you want. Minding your mind is the process of paying attention to when your presence of thinking has been switched to the "off" position because you've been triggered by someone or something in your environment.

Leaders lose their presence of mind when they are "triggered" by something in their environment such as bad news, a crisis, an obstinate colleague, and so on. One helpful DIY strategy for minding your mind is to simply list the things (or the people) that trigger intense emotions like anger, fear, or frustration in you. An even more helpful exercise is to ask someone to create their own list, and then sit down and compare notes. See how many things come to mind immediately in response to the following statement: "Nothing gets me more worked up than when people [or a specific person], _____."

Repeat this exercise several times and analyze what emotions you feel when these things trigger you. Ask yourself, Why do I get triggered by these things? How do these things threaten my ultimate purpose? Are they real threats or just perceived threats? Minding your triggers is the first step toward developing a sense of control over them so that you can be more present, make better decisions, and engage in actions that produce the results you want to achieve as a leader. Once you're comfortable with your list of triggers, consider having a conversation with a peer, your boss, or the people you lead about how these triggers impact your work together, and what each of you can commit to doing differently to support your shared purpose.

Minding your mind also means developing rituals and habits to help you take care of your mind, give it rest, and let it decompress. Our brains are wired to take shortcuts because they only have capacity for so much complex information-processing activity. Computers are much better at "always-on" processing. Humans are not. We need to stop pretending we're as good at these shallow, computational, and repetitive information-intense tasks. However, when humans use their limited cognitive capacity in combination with machines and AI insights, we can be a powerful team.

As Nobel Prize–winning psychologist Dr. Daniel Kahneman explains in his best seller, *Thinking, Fast and Slow*, our brains take shortcuts because, ultimately, they're lazy.[5] Default thinking is full of cognitive shortcuts. These shortcuts allow the human brain to call upon all of its cognitive faculties when we really need them to do deep, creative work. Interrupting default thinking, however, takes effort, for example, when you have an important decision to make.

Writing down all of the activities that energize and engage your brain, as well as all those that tax and exhaust your brain over the course of a day or week is a useful exercise. This list will help you monitor and manage your cognitive reserves so that you can start "outsourcing" shallow tasks that deplete your cognitive capacity for fully present work. Minding your mind also helps you determine when you are at your best. For example, knowing what time of day you do your best, most creative work can help you structure appointments and desk time. Do you know what times of the day you can afford to operate in default thinking mode? Do you know what time of day you need to "shut off" your brain to restore energy through sleep, exercise, meditation, or other creative activities? Minding your mind is key to managing your presence.

Struggling with Presence? Put on Your SCARF

Brain research supports the threat response that has been well documented in mind/body medical studies. Researchers use fMRI brain scans to map the portions of the brain that "light up" with electrical impulses in response to different kinds of stimuli or triggers. The research published by the Neuro-Leadership Institute[6] has found striking similarities between stressful experiences that leaders have and those found to elicit a threat response in the brain.

The application of brain science research findings to leadership development is extremely important for three reasons: (1) The types of threat responses that impact the prefrontal cortex of the brain, that is, the brain's capacity for planning, effective decision making, and complex thought, are very similar to the dynamics of leading in a 4IR environment (e.g., volatile, uncertain, complex, and ambiguous); (2) 4IR threats are going to increase in quantity, speed, and velocity over the next three to five years; and (3) the majority of organizations, governments, and professional services involved with large-scale digital transformation lack targeted approaches for helping leaders prepare for challenges and expectations that characterize the fourth industrial revolution.

Ensuring that future leaders have access to tools that help them become more mindful is extremely important. But what's more important is that high-potential leaders see their leadership role models engaged in presence-inducing, mindful, smart, connected behaviors. So, if you're a senior leader, practice presence in private for yourself. But share publicly how you maintain your leadership presence so that others know "it's okay." This is how you create a culture of presence.

Dr. David Rock of the NeuroLeadership Institute has published extensively on the SCARF model. SCARF is a very useful tool for minding your mind and for responding to the demands for greater presence in 4IR organizations. Essentially, the SCARF model represents different experiences that can trigger feelings of danger or reward. The human brain is hardwired to detect and react to these environmental triggers. The brain is constantly scanning for the five factors of the SCARF model: Status, Certainty, Autonomy, Relatedness, and Fairness. See how the triggers you listed for yourself in the previous exercise relate to the five elements of the SCARF model. Do any of your triggers pose a threat to your status as a leader? To your certainty in your purpose? Your autonomy? How about the quality of your connections with other people?

Status is your perception of your relation to people around you. You can experience a perceived loss or increase of status. Brain research shows that status is an extremely important driver or motivator of human behavior. The brain responds to perceived threats to status in the same way it does to physical pain, and to status rewards in the same way as pleasure. The digital age and the unfolding human–machine dynamics in organizations pose a fundamental threat to leaders' status as the all-powerful commanders in charge. The 4IR technological environment will also threaten the status of employees at all levels, as some will be replaced by and/or need to work alongside machines, which have a greater capacity for learning and managing complexity than humans do. These human–machine dynamics will produce enormous amounts of threat responses and anxiety that will need to be productively managed. A smart, connected response to threat is one that is mindful, present, and *responsive* rather than scattered and *reactive*. When it comes to threats to status, I recommend staying connected to your presence of thinking, feeling, and to your purpose. These are your anchors, providing you with a sense of clarity and coherence as things like titles, reporting relationships, and/or organizational structure change around you.

Certainty entails the brain's neurological response to extreme sources of comfort or discomfort. The brain craves certainty. When ambiguity exists around things like "change" in an organization, leaders' brains automatically attempt to make sense of that uncertainty. To resist change is natural because change introduces varying degrees of uncertainty and ambiguity. Resistance can take the form of emotional reactivity (i.e., lack of presence of feeling), active resistance (lack of presence of acting), or passive resistance or general dis-ease (lack of presence of thinking).

When uncertainty arises, leaders can create greater presence by antici-pating lack of presence and resistance among their followers and teams. Leaders can create greater presence and certainty amid ambiguity simply by saying something like, "When we meet with experts next month, we will know more." This simple technique can activate rewards circuitry in peoples' brains.

Providing a sense of certainty through one's leadership presence, even amid chaos, is a fundamental demand of 4IR leadership. Remember the captain's story and the TFA loop—strive to create certainty, but be authentic and trans-parent, as people can easily detect deception, which may trigger uncertainty and fear in them. "Did it seem like she was hiding something? What was she hiding? Why would she hide it?" You get the picture.

Autonomy is about feeling a degree of choice and control. When stress levels are high, leaders and employees need to know that they have choices. Empow-ering followers with a sense of control and choice, even if it's over mundane details, such as having the autonomy to decide when a group will meet, what color the new office will be painted, or where the off-site meeting will be held, will help create a sense of autonomy and choice, moving them away from a threat response state and closer toward a reward state.

Leading in a 4IR environment will provide ample opportunities for lead-ers to generate shared power and decision making, autonomous communi-ties of experience and practice, and microcultures of autonomous work teams within their organizations. For leaders to achieve these kinds of collaborative connections at work, they must release their own need for status and control. If you are struggling to let go of your need for status and control, then go back to your list of triggers. How are status and control serving your ultimate pur-pose? Is your need for status and control helping or hindering your ability to achieve your primary or secondary goals? What's even more important to you than status? Go back to your purpose.

Relatedness is about having connections with others. Threats to related-ness include not knowing others around us, perceiving others around us as different or having competing interests, or simply not knowing members of a new team. Such threats to the brain's sense of relatedness can induce a great deal of stress. I work with executives every week who feel that their team "just doesn't quite trust each other yet." They say things like, "This is a really new leadership team, and they just don't know each other that well." Threats to relatedness on leadership teams are extremely common in organizations experiencing unending, emergent change. However, leaders can minimize threats to relatedness and create more effective team functioning by identifying shared purpose and goals and creating shared vision. Building shared futures starts with relatedness and shared purpose. For some teams this takes many months. But it doesn't have to!

Improving relatedness and becoming a smart, connected team can be achieved in a few hours. Peer-to-peer dialogue (e.g., around your emotional triggers, purpose, the 5 Whats, etc.) and facilitated acceleration workshops can help teams make remarkable strides in relatedness in a very short period of time. The future of work is a team sport. Leaders must become highly skilled team builders and interpersonal connectors.

The best leadership team builders assemble teams by aligning purpose, process, people, and performance expectations. They build teams around what individual members do best, and around shared goals that advance the organization's mission and purpose. The best team builders also understand the importance of team dialogue. Tools for team dialogue will be covered in the following sections.

Fairness is the final element of the SCARF model. A perceived lack of fairness and injustice can create a threat response in people. However, when leaders understand and err on the side of transparency and fairness, they can create a feeling of safety and move followers away from feeling a threatened state and toward a reward state. One of the major threats to fairness that I see on the horizon is the perceived lack of fairness among people whose careers will be impacted by technological mega trends. For example, people who lack access to new technologies, who aren't included in the broader social conversation about the future of work, and/or who lack a sense of influence or control over what they learn are highly prone to experience a threat response to fairness in industry 4.0. This perceived lack of fairness could have a number of negative effects on organizations, from workforce planning and employee engagement to labor movements and customer distrust of governments and corporations. Leaders must start scenario planning around these possible threats to fairness and formulate plans for ensuring responsive and responsible leadership through this Great Restructuring of the world of work.

In the following, I've adapted basic principles from the NeuroLeadership's SCARF model[7] and provided recommended "thought starters" to guide leadership action in response to each element of the model. These questions will help you recognize threat responses in others, as well as in yourself, so that you can more effectively regulate and manage your mental, emotional, and behavioral responses to 4IR trends.

Table 2.1   Tool #4: SCARF Summary and Thought Starters

	Summary	Move Away from Threat	Move Toward Reward
Status	*Our perceived relative importance compared to others*	What do followers feel they are losing in terms of status?	What actions will provide followers with hope and assurance of their new status?
Certainty	*Our ability to predict the future*	How is withholding information from the team serving your purpose?	What can you say or do to create a small sense of certainty about what's to come?
Autonomy	*Our sense of control*	What power or choice are you withholding from followers?	What power of choice can you empower followers with?
Relatedness	*Our sense of knowing and connection with others*	What opportunities have we created for relationship building?	What actions would help maximize trust and relationships?
Fairness	*Our sense of equitable exchange with others*	Is this exchange equitable?	What are you willing to do to make this exchange equitable?

Use these questions when:

1. You want to better understand your own emotional triggers and threat responses.
2. You and your leadership team are making important business decisions that will impact people.
3. You are trying to understand or manage peoples' resistance to large-scale change.
4. You need to create a strategic plan to move followers toward a reward state and away from a threat response.

The first step to cultivating greater presence in your team or your organization is cultivating presence in yourself. I hope the tools and techniques in this section have helped inspire your thinking about your purpose, and how your purpose and meaning in work and in life guide your thoughts, feelings, and behavior.

In addition, I hope that you've taken away some personal insights about when your presence of thinking, feeling, and acting are challenged. By better understanding your purpose and threat response triggers, you will be well on your way to becoming a future-ready leader. In the next section, I will provide you with strategies for cultivating presence in others, which is critical for creating future-ready teams and future-ready organizations.

Cultivating Presence in Others

Leaders who build cultures of presence will more effectively adapt to 4IR mega trends, technologies, and industry disruptions. Organizations that master presence of thinking, feeling, and acting will drive innovation and productive industry disruption. A culture of presence is defined by a clear purpose and set of values that are practiced through behaviors across the organization.

Table 2.2 offers a sample of the types of values and behaviors that one might expect in a culture defined by presence of thinking, feeling, and acting.

Table 2.2   Tool #5: Sample Values and Behaviors that Define a Culture of Presence

Values	Behavioral Practices
Transparency	Candid conversations between leaders, followers, peers, and customers/constituents
Honesty	Courageous or unpopular decision making, and accountability for those decisions
Humility	Mutual respect, forgiveness, and grace
Collaboration	Mindful inclusion of others; rapid team development and redeployment
Relationships	Understanding of individuality, trust, and compassion for human struggles
Compassion	Empathic listening, patience, caring about others pain, a commitment to reducing suffering in the world
Fairness	Not treating everyone the same, but treating people in ways that they want to be treated; giving people the benefit of the doubt
Systems	Connecting parts and wholes together; discussing our interdependence; solving for underlying causes and not just presenting symptoms
Accountability	Taking responsibility and ownership for your personal actions, and supporting those around you to deliver on your shared commitments

This list is by no means comprehensive. It's not intended to be. It is intended to help leaders like you contrast the behaviors that you've observed in your organization with behaviors that demonstrate a cultural commitment to presence. Leaders often don't know what a fully present organization looks like. Mindful organizations are rare. That's because, most of the time, the people who make up an organization fail to live its values.

To illustrate this point, try this communication experiment with your leadership team. Use the values and behaviors listed in Table 2.2—or better yet, use your own organization's values—and ask your team, "Do we value these things?" Most likely you will hear a lot of "yeses" and see a lot of heads nodding. If you do, then ask your team, "Can someone share an example from the last 30 days when you saw one of these values in action?"

You'll hope to hear some great stories about transparent communication and inspiring examples of accountability. You'll hope to hear lots of compelling stories from your team about your organization's values in action. If you do, that's great! However, if you don't, ask your team the following questions, "What—if anything—would change for our employees, customers, and/or shareholders if these values were more present in our actions? Are our behaviors aligned with our purpose and values?" This conversation could take a while, but the answers to these questions will help you align your leadership team around the values that matter most. This kind of "values work" is as important at the team and organizational levels as "purpose work" is at the individual level. A culture of presence isn't possible without clear purpose and values.

Relationships and interpersonal connections are the next stop on a leader's journey to creating a future-ready culture of presence. Peer-to-peer dialogue is my preferred tool and process for cultivating presence in others. Peer-to-peer dialogue also helps you cultivate presence in yourself—what a bonus!

Peer-to-Peer Dialogue

Dialogue is one of the most powerful strategies for cultivating presence in others and future-proofing your organization against 4IR threats. Most leaders, however, don't understand what dialogue is, why it matters, how to do it, or how to scale it. True scalable dialogue involves more than basic skills training on "crucial" forms of communication—it involves mindset change at all levels of an organization.

I have been studying the concept of dialogue for 20 years, and it has become the bedrock of my system for coaching and consulting clients on building future-ready organizations. I have helped organizations leverage dialogue to overcome racial and ethnic differences, resulting in more inclusive communities and organizations. I have also used dialogue in the context of leadership development, change management, organizational design, integration resulting from mergers and acquisitions (M&A), talent and succession management, workforce planning, and large-scale culture transformation projects.

Through my extensive research and experience consulting with clients on organizational dialogue, I can confidently say that dialogue is the single greatest contributor to employee engagement, effective change management, innovative problem solving, and high performance in organizations around the world. Every organization I've worked with has benefited from improving the quality of dialogue that occurs between peers, teams, and even with customers.

Dialogue is more than mere talk between two people. The English term actually comes from the Greek work *dialogos*, where *dia* means "through" (not two) and *logos* means "the word" or the meaning of the word. When we are in dialogue with others, we are literally creating a shared reality through the meaning of the words we exchange. Consequently, dialogue and co-creation of meaning is the mechanism through which leaders can begin to build shared futures with the stakeholders that matter most to their organization's success.

Dialogue is a special approach for making meaning and creating understanding with others through our words and deeds. Dialogue is more about the **process** of conversation and meaning making rather than the **outcome** of "passing" your meaning onto others. It's like playing a game with the child. The fun is in playing, not in who wins. In leadership communication, as in business operations, better processes lead to better outcomes. Dialogue is a better communication process for building a shared future and future-ready teams and organizations.

Think of dialogue this way: if everyday conversations are about **transactional exchanges**, then dialogue is about **transformational emergence**. In everyday communications, we exchange meaning (or at least attempt to) in a transactional way. For example, when I order my Starbucks coffee, I tell the cashier, "I would like a Grande Flat White." He then asks whether or not I would like anything else, to which I reply, "No thank you." The cashier then tells me the drink will be a certain price, and I pay him that amount.

Transactional communication is simple, right? Meaning was exchanged, and the transaction was completed. This was a highly pragmatic exchange in which everyone got the outcome that they expected. I got my coffee and Mr. Starbucks got his five bucks. In this example, meaning was passed back and forth like a basketball. We didn't create new, emergent meaning because we didn't need to. We simply had a transactional exchange of meaning.

What makes these transactional exchanges work is that both people know what to expect and both are well grounded in the situational demands of the context. Ironically, I often see some of the simplest transactions of meaning in large organizations fail—even with advanced communications technologies. Leaders often feel like they've "ordered a coffee" from their team, and then they get frustrated when the team delivers a "banana split." I'm sure you've experienced this as a leader. Just because you've said something does not mean that you've communicated anything at all!

Dialogue, on the other hand, is completely different than transactional exchanges like the Starbucks example. Dialogue is the preferred communication process for when the people involved aren't clear about the demands of the situational context, and when they don't know what to expect from the process or when things change. Dialogue involves presence of thinking, feeling, and acting *together in conversation*. If I were asked, "How do future-ready organizations communicate?" my answer would be that they're great at dialogue. The point of dialogue is to make sense of the world *with* (not for) the people involved in the process. Given the volatility and uncertainty of the 4IR economy and workplace, effective dialogue is essential for surviving and thriving in the new world of work.

Dialogue requires a different mindset about communication as well as different communication skills. The mindset is one of systems complexity, presence, and an openness to emergence. Dialogue involves deep listening and responsive questioning. Dialogue requires that the people involved have deep self-awareness (presence of thinking) about their own mindset and triggers so that they can recognize how statements, ideas, and questions make them feel. How they make others feel. And how a group or team might "stay present with" an idea to fully hold it up and explore its viability. Dialogue, in words of quantum physicist and dialogue theorist, David Bohm is about *thinking together* with others.[8] You can't get much smarter or more connected than when you're thinking together with others. Well, I suppose that would depend upon who you're thinking together with, but you get the point!

The good news is that a dialogue-driven mindset can be learned and developed. In addition, the communication skills necessary for dialogue can be learned. I've had great success teaching university students and senior executives alike the dialogue state of mind and skills set for the last 20 years. Good peer-to-peer dialogue starts with finding the right time and space for dialogue. This allows for people to approach the dialogue with greater presence of thinking.

Dialogue requires presence of thinking and feeling because the people involved have to practice responding to triggering statements and events with curiosity and compassion, not defensiveness and threat responses—as these types of responses only escalate dysfunction and unproductive outcomes. Dialogue demands patience and time. People must take time to set ground rules for engagement. For example, participants must agree to lead with inquiry to seek understanding, as opposed to simply leading with advocacy for their point of view. In other words, you have to ask questions, not just wait for your turn to talk and convince others of your status and intelligence.

Finally, dialogue requires that you begin the process of communication before knowing exactly what the outcome might be. Dialogue is emergent. You might go into a dialogue thinking that you're going to get that coffee that you ordered but might find yourself delighted by the banana split that emerges from the process. Creating time and space for dialogue is the first thing that comes to mind when I hear leaders say things like, "We have to slow down to go fast." In my mind, this means we need to improve dialogue before we can take action. We need to have greater presence of thinking and feeling together before we collectively take action. However, many of these same leaders get "stuck" by having the same quality of transactional exchanges while they've "slowed down" because they don't have a dialogue-driven mindset or skills set. There is no better way to engage when a team is attempting to future-proof their organization than with dialogue. Dialogue is the only way that people can truly achieve shared understanding or build innovative shared futures together. Good dialogue is the essence of emergent design and shared future building.

Here's a case in point: I was working with a leadership team a few months ago on the topic of "the future of work." An executive from a $13 billion enterprise convened his top 20 leaders for two days offsite. The purpose of the meeting was to explore big ideas about the future of work, process those ideas, and walk out with a high-level plan for preparing the organization for the types of 4IR mega trends most relevant for creating value across their more than 30,000-employee enterprise.

On the morning of the first day, my team and I grounded everyone in our purpose for the meeting and reviewed the desired outcomes that we would be working toward. We then built shared understanding around the five mega trends are that are impacting the client's industry and business. We hosted several industry experts who presented on topics such as artificial intelligence, robotics, and workforce planning. Following my presentation, my team facilitated table discussions around the questions, "Which of the five mega trends is most likely to disrupt your business?" and "What early signs do you see that these changes are already taking place?"

I observed the people in the room during the morning presentations and the table discussions. I was paying close attention to the expressions on their faces and other nonverbal behaviors. The expressions on their faces in response to the data and ideas that were shared about the future of work were that of wonderment, intrigue, curiosity, and bewilderment. Even though this group asked great questions of each of the presenters, my sense was that the "Q&A" time was insufficient for building a common understanding among the group.

Without a common understanding of these future work trends and ideas, I knew that the group would struggle with the "so what" (implications) and "now what" (action planning) portions of our afternoon session. Therefore, I knew I had to do something unplanned and emergent with the group to help them achieve the outcomes that they wanted (and that my client—their senior executive) wanted out of the meeting (i.e., a plan to prepare for the future of work).

After they'd completed the "big ideas" session in the morning, I proposed to the client an emergent design framework for deepening dialogue that had begun during the Q&A. He agreed with my revisions to the agenda. His response was, "This sounds great. I love emergent design." Spoken like a true future-ready leader!

The framework that I put into play is called *Open Space*. According to openspaceworld.org, "Open Space has been a daring and marvelous exploration of the vastness and the urgency of personal and organizational transformation. For others, it's just an exceedingly effective, and efficient, meeting methodology."[9] My gut told me that this dialogue methodology would be helpful for deepening the team's understanding of fourth industrial revolution mega trends, and for exploring what they meant for their company's future work strategy.

My experience facilitating group dialogue using an Open Space meeting structure has been very positive. The rules of Open Space are simple:

1. *Whoever comes are the right people.*
2. *Whenever it starts is the right time.*
3. *Wherever it is, is the right place.*
4. *Whatever happens is the only thing that could have; be prepared to be surprised!*
5. *When it's over, it's over (within this session).*

There is also one final rule known as the "law of mobility" or the "law of two feet." According to Open Space's creator, Harrison Owen, the law of mobility states that people can move around. If at any time during the dialogue someone feels they aren't learning or contributing, they can move to another small group and take up the conversation there. No hard feelings.

What happened next with our large group could only be described as unanticipated, organized chaos—and remarkable. I simply introduced the Open Space rules, asked that the groups be ready to share a short presentation of what they talked about, and then stepped out of their way, drank coffee, and listened. In the hour and 15 minutes that followed, the group of 40 leaders self-organized into five different groups, addressing topics ranging from customer experience, employee experience, automated work design and analytics, to building a talent incubator.

Once each group was finished, they presented their brief summary of what they had talked about. Through the small-group dialogues, amazing ideas emerged and shared futures began to take shape. The collective understanding of what the future of work meant in this company's industry, market, and context was more fully realized and shared across the team of leaders.

Through the Open Space dialogues, the group collectively identified which elements of future-proofing their workforce were most important and meaningful to them. This set the team up for an even more productive afternoon and second day of strategic planning around the future of work. What's more, the Open Space dialogue was efficient. After the brainstorming and reports, we were actually seven minutes ahead of schedule on our agenda and early for lunch! Lesson learned: with the right people, right time, and right facilitation, emergent dialogue can be more creative, effective, and efficient than conventional strategic planning structures.

Wanted: More Deep Dialogue

I perform a lot of diagnostic assessments on organizations to help determine their future-readiness. In almost all cases, "communication" is either an explicit barrier to or an opportunity for improved team effectiveness. Whether I ask 10 employees or 300,000 employees, "If you could change one thing to improve your organization's effectiveness, what would it be?," *communication* (or some version thereof, such as improved management, clarity of leadership vision, more communication, better communication from senior leaders, etc.) is always in the top three most frequently cited items. The bottom line: everyone wants deeper dialogue. Leaders want it, managers want it, and employees want it. So, what do we have to do to start having deeper dialogue?

Nine times out of ten, organizational communication barriers are related to *poor quality* of dialogue, not to the organization's ability to engage in back-and-forth transactional communication. Although, you wouldn't believe how often I see organizations that still have massive barriers to basic transactional communication. I see of lot of these kinds of communication barriers in health care organizations too, which is terrifying! These communication barriers include things like the following:

- Frontline employees don't have an email address;
- Leaders don't have a technological platform to get the same message to all employees at the same time;
- Internal policies and procedures restrict basic information use and information sharing;
- Regulatory policies restrict who has access to information and the means of sharing that information with those who need to have it—this is often for good security reasons; and
- The company lacks effective meeting structures and coordination to ensure that messages are "cascaded" from the top down or from the corporate office (center) to the field (out) in an efficient and effective manner.

Assuming that organizations solve for these basic "transactional communication" barriers, the next muscle to build is high-quality peer-to-peer dialogue. High-quality dialogue requires a dialogue strategy, a common dialogue process that is used across the organization, a change-management plan for evolving everyone's mindset about dialogue (i.e., a business case for dialogue, communication, education, and analytics to demonstrate return on investment).

By implementing a strategic plan for **improving peer-to-peer dialogue**, many of the common barriers to high-quality dialogue can be overcome. These barriers include lack of trust, perceived lack of transparency, lack of inclusion, low-quality management, need for basic communication skills, and need for advanced dialogue skills like facilitation, sense making, advanced inquiry, and so on. Not only do leaders and employees crave deeper dialogue, the technological and social complexity of the future demand deeper dialogue in organizations and communities. Leaders cannot solve today's complex problems with tweets and "messaging" an issue. Solving today's problems requires establishing deep communication and feedback, learning from mistakes, creating a shared mindset, aligning on a direction, building consensus, and creating shared futures.

It seems, however, that we're living in an era where the public has given ultimate authority to the loudest voice and/or the so-called experts with the biggest microphone, Twitter following, and so on. This is true in government, politics, corporations led by "celebrity" executives, and all sorts of industry associations. The "influencers" of these tribes aren't those with the strongest science, rational arguments, or even logical claims to truth; the winners are the best marketers.

I discovered this firsthand when I was searching for a publisher for the book you're reading now. Nearly every editor who reviewed it gave praise for the forward-thinking concepts, the practical tools, and the well-researched claims; however, the first 10 publishers—for whom I have a great deal of respect—rejected it. Ironically, they also said that they were looking forward to reading the book when it came out. Interestingly, 8 of those 10 explicitly stated that the reason they didn't want to publish the book was because of the size of my social media platform. This is the world we live in. Twitter followers—many of which are bots anyway—are now the gatekeepers of which ideas get shared in the world from our most credible sources of media. This is truly a new reality of "artificial news." Consequently, future-ready organizations and communities must get better at creating shared meaning, separating fact from fiction, and creating cultures where the best ideas win and are implemented.

Despite my firsthand experiences of ineffective dialogue and mindsets that seem more motivated by profit rather than purpose, I still believe that people want to read and discuss ideas that are based in science, fact, and expert opinion. I know that people want to have a deeper dialogue about the issues of our day and how those issues will impact the future of work and life. Many leaders seem to have lost their way in our highly distracted world of tweets, likes, and "shares." What seemed like powerful media for amplifying the best of the human condition, as Clay Shirky warned in his 2008 book, *Here Comes Everybody: The Power of Organizing without Organizations,* social media have proven to be equally as effective in amplifying the worst of the human condition.[10]

For leaders who want to have a deeper conversation about the future of their organization, my analysis of organizations suggests that there is a huge appetite for deep dialogue among constituents of all sorts (e.g., employees, students, customers, and voters). Designing impactful dialogue requires a mutual partnership with people. It requires a shared physical space or a richly simulated virtual or augmented space. Dialogue requires a willingness to create shared meaning, outcomes, and the capacity to coordinate action together in the direction of a shared purpose. Impactful dialogue is essential for innovation and emergent design thinking and creative problem solving. The process isn't always easy or comfortable. In fact, dialogue can be pretty darn "messy," but the outcomes are usually surprising, rewarding, and sometimes pure artistry!

If you want more impactful dialogue in your organization, you have to remember one thing: *in most organizations, impactful dialogue usually only happens by accident.* But it doesn't have to. By being more intentional about when, where, how, and with whom you design future-focused dialogues using tools like Open Space and the ones I've outlined in this section, your team and organization can achieve better outcomes. Forward-looking dialogue can help you prepare for the future of work faster and build collaborative advantage over your competition.

Table 2.3 outlines the core points of contrast between default conversation and smart, connected dialogue.

Table 2.3 Tool #6: Contrasting Qualities of Default Communication versus Smart, Connected Dialogue

Default Communication	Smart, Connected Dialogue
Transactional	*Transformational*
Separate	*Mutual*
Roles	*Souls*
Pragmatic	*Creative*
Default (mindless)	*Design (present)*
Predictable	*Emergent*
Rational	*Rational and Emotional*
Guarded	*Trusting and Open*

Smart, connected dialogue is one of the best ways to cultivate a culture of presence because people have to be "all in" together. A person who isn't fully present and participating in the process sticks out. During smart, connected dialogue, people acknowledge other participants' contributions. Participants take the time to respond to others' contributions with curiosity, as opposed to waiting for their turn to make a statement.

What makes this quality of conversation "smart" is that people participate and lead with questions (inquiry) rather than judgments, assertions, or statements of certainty (advocacy). This spirit of inquiry leads teams to a deeper understanding of the issues at hand. A spirit of inquiry does not mean that people cannot make statements or advocate for their point of view. When facts, assertions, and statements of "fact" would help advance the team's understanding or ability to move forward together, then participants are obligated to bring them to the table.

What makes this type of conversation "connected" is that participants approach the conversation with shared purpose and goals that connect them. Building belief and trust around shared outcomes can take time and a commitment to multiple smart, connected dialogues, but only through this process can reconciliation, shared vision, and future-focused innovation occur. When smart, connected conversations are at their best, there is a high degree of trust among participants, which reinforces the connections among them.

Through the dialogue process, participants strengthen their connections around a shared vision and purpose, which unifies the conversation and allows people to call one another out on how their contributions are helping or hurting the group's progress. The best part about smart, connected dialogue—despite the fact that every year it seems leaders and their organizations have less and less time for deep communication—is that the skills can be learned. Teams that have not experienced deep communication about a topic like the future of their organization often need to refresh their skills around deep listening, inquiry versus advocacy, mutuality, and feedback. However, with a little education on the ground rules and some supportive facilitation to get the ball rolling, it usually doesn't take long before someone says, "This is really great. Why don't we talk like this more often?" From what I've observed, the answer to this question is because people are "too busy" with schedules like Gerry's. Or perhaps they're just busy building their Twitter following for their next book!

Although smart, connected dialogues usually require face-to-face meeting, technology is improving our human ability to engage one another in smarter and more deeply connected ways. Technology allows us to shift the *when* and *where* of deep conversations in ways that founding dialogue theorists and practitioners like Martin Buber, Carl Rogers, Mikhail Bakhtin, David Bohm, and even dialogue advocates like Dr. Martin Luther King Jr. could have never imagined. For example, Ray Dalio, Bridgewater's CEO and one of the 100 wealthiest people in the world, tells a story in his 2017 TED Talk about the "Dot Collector."[11] This is a tool that Bridgewater uses to build a culture of transparency, wherein people can rate the quality of one another's contributors to a team discussion.

Dalio explains, "As the meeting [on President Trump's impact on the economy] began, a researcher named Jen rated me a three—in other words, badly—for not showing a good balance of open-mindedness and assertiveness." This technology allows for open and honest feedback and is an attempt to flatten hierarchy so that the best ideas in the company win.

Dot Collector and other such tools are by no means a panacea for having smart, connected dialogue. Such tools do, however, change the nature of communication in organizations. When used with great thought, care, and skill, communication tools like the Dot Collector can help make teams more aware of how open and honest they are being with one another. They can build trust and shared understanding of things like purpose and possibilities in the organization. But you can be sure that a culture of transparency requires more than a rating system and "scores" on the comments that people make in meetings. It requires leaders who have presence of thinking, feeling, and behaving, and who are committed to creating a culture of presence within their organization through smart, connected dialogue.

Leaders should explore the principles of smart, connected dialogue listed previously, and experiment with the tools in this section for the following applications:

- Creative problem solving
- Complex decisions making
- Conflict management and handling other "crucial" issues
- Design thinking and innovation
- Collective visioning
- Strategic planning
- Future-proofing strategy sessions
- Creating shared futures

I find two tools particularly beneficial for helping leaders design more impactful dialogue. And, no, these tools won't require investment in a custom "Dot Collector." I pull these tools out of my "future of work tool kit" more and more these days, as the pace and velocity of change increase in the organizations I advise. My guess is that these tools will become even more valuable as the scope, speed, and complexity of 4IR problems increase. So, hold onto them, and share them with others in your organization to help them start smarter, more connected dialogue.

The first tool for impactful dialogue is the peer-to-peer dialogue guide (P2PDG). You should use the P2PDG when:

- Cultivating and practicing presence yourself
- Teaching dialogue skills and helping others develop presence
- Introducing dialogue to your team to support innovation
- Experimenting with the dialogue process to resolve conflict
- Learning how to apply dialogue to group problem-solving tasks
- Establishing trusting relationships with peers

The P2PDG provides a set of guidelines and questions that you can use to architect a one-on-one dialogue with a peer or impactful dialogue with a small group. The P2PDG questions are thought starters and conversation catalysts. These questions create interpersonal value in that they help team members learn more about one another, sharpen their communication skills, and become more present in the conversation. The P2PDG guidelines can be applied to "real-time" dialogues around 4IR workplace challenges that your company is facing, or can be integrated into a larger strategic planning meeting or into an organization's future-proofing strategy of workforce learning and development (more on this in an upcoming chapter).

In the left-hand column of the P2PDG are questions for you to ask your key partners. In the right-hand column of the guide are questions for you to reflect on in the moment and/or following the dialogue with a peer. As you use the P2PDG, practice being present with the person with whom you're communicating. Observe what your business partner is saying. And take note of what's not being said. You'll often find insight, wisdom, and opportunities for improving trust with your partners by paying attention to and being responsive to what they are NOT saying. Our ears our pretty tuned into what people say, but many leaders struggle with hearing what's not being said. Practice listening to the spaces between another person's words.

While you use the P2PDG, remember the insights you learned from the tools outlined earlier in this section, like practicing self-awareness and relationship management (emotional intelligence), minding your mind, watching out for SCARF threat response triggers in yourself and in your partner, and working toward a reward state through your tone of voice, eye contact, and body language. Communicate care, compassion, or at a minimum that you give a darn about this person by being fully present. For some, this might be a challenge. Having facilitated this exercise with hundreds of leaders, I know you'll be surprised what emerges from the dialogue!

Table 2.4 Tool #7: Peer-to-Peer Dialogue Guide: Guidelines, Leadership Competencies, and Conversation Starters

Topic	Questions for Peer	Questions for You to Think About
Knowing Others		
Work	What events in your life brought you to this career? What are you most passionate about in your career?	What does this person value? What motivates them?
Life	What would you do more of, if you woke up tomorrow and chose to leave this work behind?	How aligned are this person's personal and professional passions?
Partnership	In what areas do we work best together?	What contributes to the strength of your partnership with this person?
	How might we work even better together?	Does this person think the work you partner on is well aligned?
Focusing on Followers	What do our followers need more of from us? What do our followers need less of from us?	How well are we meeting the needs of our followers?
Enhancing Collaboration	When are our respective teams at their best? What would help our teams be even more effective together?	What can we do as leaders to influence better team collaboration?
Wowing Customers/ Constituents	What do our customers appreciate most about us? What would need to exist for us to delight our customers or constituents even more?	How customer- or constituent-centric are we?
Achieving Goals & Action	What priorities are most important to our organization or institution?	What am I willing to do to support my peer, my team, my organization to achieve these goals?
	What priorities are most important to you personally? What actions can we take to move closer to achieving these goals together?	What support do I need from her, him, or others?
Action Guide	Who, does what, by when?	

Guidelines for Using the Peer-to-Peer Dialogue Guide

1. This is a guide to designing impactful peer-to-peer dialogue. It's not a script for dialogue. Go where the conversation takes you whenever you are moved to do so.

2. These questions are thought and conversation starters that can be used to practice leadership dialogue skills.

3. Practicing dialogue skills with this tool will help you, your peers, and/or your team develop dialogue competencies around presence, listening, and communicating.

4. Transfer questions from this dialogue guide that you find helpful to different dialogue situations (e.g., innovation, problem solving, visioning, strategic planning, future-proofing, and/or conflict resolution situations).

5. Familiarize yourself with both the questions that you will ask your peer or team ("Questions for Peer"), and also get familiar with the "Questions for You to Think About" just before you meet with your business partner. The latter might make good follow-up or probing questions, which could lead to even more impactful conversation.

Fourth industrial revolution challenges demand that leaders:

1. Start a different kind and quality of conversation (dialogue) among their peers and followers.

2. Respect and accept their followers as unique individuals, not as roles or "assets" to be leveraged.

3. Are mindful of how they show up and participate in 1:1 and team dialogue.

4. Influence with inquiry and curiosity, not advocacy or certainty.

5. Are open to the possibility of having their thinking, feeling, or actions influenced by another.

These questions and points of reflection are extremely powerful tools for building relationships with peers and with teams. As you practice asking the kinds of questions in the P2PDG, you will develop dialogue competencies and communication skills like active listening, greater presence of thinking, feeling, and acting with your peers. The P2PDG is one tool to help you develop leadership presence in ways that work for you, your peers, and your team.

In sum, future-ready leadership requires a different kind of human-to-human connection. These relationships are more open, engaging, transparent, and trusting. They involve a deeper level of listening to understand how others are thinking, feeling, and what is driving their actions. For some leaders, their presence of thinking, feeling, and communicating is highly effective and it's just who they are as human beings. For the rest of us, like Gerry, being present and leading with purpose can be a challenge. As leaders of today and leaders of tomorrow, we have to work hard to develop the self-awareness, self-regulation, and dialogue skills that the changes of the fourth industrial revolution demands of us.

Chapter Summary

1. Effective 4IR leadership demands that leaders develop greater presence of thinking, feeling, and acting. Leaders need a repertoire of tools, practice, and support using these new techniques to improve their leadership presence and engage in more critical, systemic, culturally agile, and innovative ways of thinking.

2. This chapter introduced seven tools to help you develop your leadership presence: the Thought, Feeling, Action (TFA) Loop; the Purpose, Goals, Tasks Hierarchy; Switch Your Leadership Presence on by Minding Your Mind; SCARF Summary and Thought Starters; Sample Values and Behaviors that Define a Culture of Presence (Table 2.2); Contrasting Qualities of Default Communication versus Smart, Connected Dialogue (Table 2.3); and the Peer-to-Peer Dialogue Guide.

3. Creating a culture of presence begins with presence and dialogue.

Recommended Actions

1. Switch on your presence. Use the TFA loop to identify your leadership presence strengths and weaknesses.

2. Stop having default conversations. Start having smart, connected dialogues.

3. Use the P2PDG or an Open Space meeting structure as soon as possible to improve collaboration with your key business partners.

CHAPTER THREE

Agility

"In conclusion, the only limiting factor of our Acme 2025 digital transformation strategy is our ability to attract, hire, and retain qualified employees with the right technical skills. We are confident in our plan for this year and have already begun implementation to address this challenge."

Jane's presentation was met with silence and looks of concern across the Acme boardroom. As chief people officer for Acme, Jane felt a personal sense of responsibility for her company's dismal fourth quarter 2020 earnings. This was more than an uptick in turnover or poor succession planning, the gap between the talent needed to grow and the talent available was severely limiting Acme's growth.

As she left the boardroom, concerned for her own job security, Jane thought to herself, Who would have thought back in 2018 that so many robotics engineers, game designers, security specialists, and data scientists would have been needed to grow a retail grocery store chain? Who would have thought that the grocery industry would have changed so fast in the area of mixed reality? The future of grocery just exploded after Amazon acquired Whole Foods Market.

The digital transformation of retail—and many other industries—is about dreaming up new and differentiated customer experiences. The experience economy connects brands, products, and people. These experiences differentiate and will continue to create convenience and loyalty, but they require new digital platforms, which means people with new talent. At the same time, automation, robotics, digitization, and AI are being leveraged to increase scale and efficiencies across the entire value chain.

Balancing these risks and rewards is critical for long-term relevance and sustainability. Leaders must exercise a future-focused mindset and act with speed and agility to create future-ready organizations. As Jane's scenario depicts, agility requires anticipation and boldly going where other organizations are only toying with the idea of going. Most leaders have a hard time envisioning what the next quarter will look like, let alone the next three to five years. However, if keeping your organization relevant to your customers in the fourth industrial revolution is important to you, then becoming more responsive and agile in the face of high-velocity change is something that you had better build for now. Otherwise, you might end up in the same position that Jane finds herself in.

If high-velocity change is the challenge of our times, then developing an agile workforce and leadership team is a critical ingredient to the solution. To be clear, I'm using the word *agile* to describe a leadership mindset and set of behaviors that leaders need to cultivate within their organization to respond to high-velocity change. I realize there is an established body of literature, training, and certifications in the agile method of software development and workflow. Tools like scrums and sprints are complementary to the type of agility that I argue for in this chapter. However, this chapter is focused on the tools and techniques for developing an *agile mindset*, culture, and aligning your organization around things like risk, failure, love of learning, rapid prototyping, and so on, all of which are required for a future-ready organization.

Before proceeding, it's worth noting the case study at the beginning of this chapter. The velocity of change that Jane and Acme found themselves getting steamrolled by provides a glimpse of what nearly every major industry in the global economy will face over the next decade. Jane's 4IR-inspired anecdote is intended to be a wake-up call to executive teams and human resource professionals everywhere. The future of work is now, and organizations are not future-ready.

Future-Ready Leaders Are Responsive and Agile

Velocity is a defining characteristic of the fourth industrial revolution, and it distinguishes 4IR from the previous three industrial eras of our modern economy. The world of technology and business are changing faster than ever before. Our technological capability has already outpaced our human ability to harness and make decisions about appropriate uses of technology, security, privacy, and so on.

The studies I share in this section show that the growth potential of the global economy will be limited, not by too little technology or capacity for growth, but by too few people with the right skills to collaborate and drive responsible and responsive growth. As the case study of Jane and Acme grocery illustrates, game design, augmented customer experiences, and security are the technical skills of the day and for the foreseeable future. However, these technical skills must be balanced with an ability to communicate, collaborate, share status and power, and relate across differences in culture and mindset. As one of my clients put it, "We're going to need a lot more business-savvy 'quants' [quantitative analysts] who have a personality and can get along with other people." I thought this comment perfectly named the challenge of the day: How do future-ready organizations build or "buy" such talent to keep pace with competition and growth?

Finding employees who have the critical and analytical thinking skills of engineers or software developers, but who also understand the business and are socially adept is a tall order. Finding this kind of talent is like finding unicorns—they're almost mythical creatures. But many companies are telling me the same thing: they want "social quants" to drive their businesses forward. These future-focused leaders believe that talent will fuel their response to fourth industrial changes in their respective markets and industries.

One implication of companies all "hunting" for unicorns at the same time is that there isn't enough of this rare talent combination around to fuel global organizational responsiveness. At least not yet. This means companies must become better at creating more attractive and inclusive workplaces for a new wave of global talent. In addition, organizations of all kinds will have to start having serious conversations about how to up-skill their current workforce in planning for a different set of highly technical and collaborative tasks to be carried out as high-velocity technological change transforms the structure and content of the jobs. Future-ready leaders have strategies in place to both "build" and "buy" the unicorns (social quants) that they need to drive their organizations forward into the future.

As evidence of this, I'll refer back to World Economic Forum's founder and chairman, Klaus Schwab. In the introduction to his 2017 book on the fourth industrial revolution, he states, "Contrary to the previous industrial revolutions, this one is evolving at an exponential rather than linear pace." Schwab's claim is supported by convincing data from fields like information technology and economics. In terms of *pace* of change, 4IR is a different animal than the information age of the third industrial revolution. The *pace* and *direction* of 4IR changes create a unique set of job demands for leaders and the organizations they lead.

Smart, connected leadership sounds like a marketing term. But it's not. It's an accurate description of the required mindset, disposition, behaviors, and capabilities that leaders need for success in this high-velocity change environment. A smarter, more connected mindset about the future is what will enable leaders to build more fluid, flexible, and agile workplaces.

A smart, connected mindset is a way of observing, and responding to, the changes in business, society, and technology. Leaders' mindsets serve as filters for what leaders see, feel, and do. Although a smart, connected mindset is not the only ingredient for agile, future-ready organizations, it's an important place to start. Just as creating a culture of presence in your organization starts with individual leaders, so too does creating a responsive and agile organization.

Leaders who have a smart, connected mindset are forward-looking and responsive. They have their eyes on the road ahead, and they're not putting the organization at risk by looking out their rearview mirror. Smart, connected leaders pay attention to what's on the horizon, and they have the ability change direction quickly in response to observations, insights, and implications they see. This is where agility comes in. Smart, connected leaders pay attention to signs and signals that others ignore or simply don't see. In some cases, these signals come from customers, members of their community, students, governments, competitors, adjacent organizations, agencies, and so on.

The ability to see these signs and make sense of them is where the "connected" piece of the phrase *smart, connected* comes in. Connected leaders aren't just tied to social media or a technology device. Connected is an adjective I use to describe a leader's ability to make sense of the relationships between things (e.g., people, events, trends, institutions, technologies, and so forth). Connected leaders are systems thinkers. They see cycles of influence amid the chaos. They perceive patterns where others only see randomness. Connected leaders don't just "watch"; they observe.

People who don't know how to observe will struggle to see the patterns and connections between seemingly disparate marketplace occurrences. Moreover, if a leader cannot observe and make sense of connections (i.e., seeing threats, opportunities, or possibilities), then that leader will not be in a position to move in an agile fashion and will be blindsided by high-velocity change. Observation is a different skill than "looking" or simply "seeing." Observation requires a different way of seeing that intentionally makes the familiar look strange. Observation requires looking at things with a childlike curiosity or what some have called a beginner's mind. As the root of an agile response, observation requires conceptual thinking, critical thinking, speculative ("what if?") thinking, creativity, and dialogue with colleagues to make sense of relationships one's observing.

This section will provide you with frameworks for observing change differently, to help you become smarter and more connected. I want to help leaders become more agile with the tools and lessons in this section so that they can create more future-ready organizations. First, I will teach you the three elements of high-velocity change: speed, direction, and acceleration. This framework will help you observe change in a new light, and make sense of how you, as a leader, need to formulate your agile response. Next, I will provide you with tools for preparing your organization for high-velocity change. These tools will help you ready your team and workforce for making agile responses to high-velocity change.

What Is Velocity?

Velocity is a term from classical mechanics. Velocity specifies an object's **speed** and **direction** as a function of **time**. If an object experiences a change of direction, speed, or both, then the object is undergoing an **acceleration** and, hence, a change in velocity. These three variables—*speed*, *direction*, and *acceleration*—are essential for understanding how to increase your organization's agility amid high-velocity change.

Velocity is different than speed because it is a vector quantity that requires both magnitude and direction. In other words, 10 miles an hour is a measure of speed, and 10 miles an hour east is a vector quantity. Don't worry, I'm not going to go any farther into the physics of velocity, as that's way beyond my expertise. But I will continue to use velocity as a metaphor for illustrating why 4IR requires an agile mindset about what's changing in the world of work, and what leaders in your organization need to do differently to respond.

Just as the velocity of a race car driver going around a racetrack requires a different style of driving than taking a Prius around a city street corner, the velocity of the 4IR environment requires a different kind of leadership style. Speed, direction, and acceleration change the nature of leadership in a 4IR environment.

Fourth industrial revolution risks are a prime example of high-velocity change. Risks, such as a new low-priced competitor in your market, differentiated on technology and automation, could pose a threat to your organization from multiple directions (e.g., a start-up, a pilot project from an existing competitor, an emerging market, etc.). A future-ready response to this threat requires agility. Your leadership team, ideally, would have spotted the competitive technology and automation opportunity before the emerging threat. However, assuming they didn't (because they were looking out the rearview mirror), your team would need to quickly decide whether to allocate resources to mitigate this risk, how many resources, and with what level of prioritization given other organizational commitments. As this simple example illustrates, leaders need to know the rules of the road in agile 4IR organizations, and they need to understand how to become a more agile decision-making machine to avoid a crash in the face of disruption.

As with all the leadership requirements of the 4IR, there is good scientific evidence to warrant developing leadership skills around the high-velocity changes that will help you future-proof your organization against 4IR disruptions. The leadership skills in this section won't stop the disruptions from occurring, nor will they inoculate your organization against disruption. These skills will, however, prepare the human decision makers in your organization to make more agile responses to the types of high-velocity changes impacting workplaces, institutions, and government organizations.

Some Sobering Statistics about Workplace Disruption

Earlier I mentioned that organizational agility in the fourth industrial revolution will depend on attracting, developing, and retaining great talent. Before we proceed, I want to provide you with a quick "hit list" of research findings on the changing nature of the global workforce, and how that might impact your responsibility of becoming a more agile leader. These statistics come from some pretty credible sources, and they paint a compelling picture that high-velocity change is most certainly upon us.

These sobering statistics serve two purposes. First, they prove that the Acme case study at the beginning of this chapter is not only possible, but also probable. Most of the organizations I consult with already have many roles for which they have inadequate talent to fill them. Second, they offer you a chance to reflect and to start conversations with your team, your board of directors, your heads of human resources, and educators within your community about what this means for your organization and community. If the health of our organizations and institutions depends on a thriving, well-skilled, and agile workforce, then we as leaders should start workforce planning now, and not when it's more convenient. Today's convenience is tomorrow's crisis.

Fact #1: U.S. Workers Know Their Jobs Aren't Robot-Proof

A 2017 Gallup poll found that "about one in four U.S. workers (26%) say it is at least somewhat likely that their job will be eliminated by new technology, automation, artificial intelligence or robots within the next 20 years. About one in eight workers (13%) say this will happen within the next five years."[1]

Fact #2: Nearly Half of U.S. Jobs Are Highly Likely to Become Computerized

A 2013 study from the Oxford Martin School predicted that as much as 47 percent of U.S. jobs were highly likely to become computerized between 2020 and 2030.[2] A revised 2017 study, conducted in partnership with Citi, predicts as many as 80 percent of jobs in retail, transportation, and logistics are susceptible to automation, along with 63 percent of sales occupations.[3]

Fact #3: Artificial Intelligence Is Diagnosing Cancer

IBM's Watson treatment recommendations were highly concordant with recommendations made by the Manipal Multidisciplinary Tumor Board in Bangalore, India. Recommendations were concordant in 96.4 percent of lung, 81.0 percent of colon, and 92.7 percent of rectal cancer cases.[4] One implication of this fact is that humans will need to learn to optimize their productivity by developing strong critical thinking and analytical skills to work alongside machines. Machines can already answer questions faster and more accurately than humans. However, if humans don't know the right questions to ask, then both human and machine intelligence will be suboptimized or wasted.

Fact #4: Humans Are Old . . . Who Needs Them!?!?

By 2035, one in five people (20 percent of the population) will be 65 or older. These demographic changes will facilitate a "mass exodus" of talent from the workforce but will also create new leadership opportunities for millennials. The type of connections and working relationships that older knowledge workers and leaders have with employers will have to radically change to support part-time, gig, and consulting arrangements that are mutually beneficial for individuals transitioning into retirement and the organizations who need their knowledge and skills.

Fact #5: No Talent, No Growth

Almost half of U.S. and German companies surveyed by a 2017 Boston Consulting Group (BCG) study cited "the lack of qualified employees" as the greatest constraint to a full digital transformation. Therefore, growth will require more unicorns like those I've previously described as social quants.[5]

Fact #6: Colleges Might Be on to Something

Many colleges now offer majors that didn't exist five years ago (e.g., game design, data science, and cybersecurity). Ironically, industry leaders still frequently lament that university undergraduates lack the critical thinking, writing, intercultural, and communication skills necessary for "hitting the ground running" in their roles.[6]

Fact #7: Cybersecurity Is in a Crisis

According to *Peninsula Press*'s analysis of numbers from the Bureau of Labor Statistics, more than 209,000 cybersecurity jobs in the United States are unfilled, and postings are up 74 percent over the past five years. This project of the Stanford University Journalism Program also shows that the demand for information security professionals is expected to grow by 53 percent through 2018.[7]

Fact #8: Talent Supply Will Remain in a Crisis

According to the World Economic Forum, "by 2020, more than a third of the desired core skill sets of most occupations will be comprised of skills that are not yet considered crucial to the job today." In other words, technology is changing so fast that the only certainty is that skills for success will radically change—everyone is going to have to update their knowledge, skills, and capabilities.[8]

Fact #9: Yep, Looks like We're in Big Trouble

BCG also predicts a global workforce crisis within the next 15 years with labor deficits in the world's 15 largest economies, which make up 70 percent of global gross domestic product (GDP). Implication: this global labor deficit will impact almost every large multinational company. This is not to say that the predicted workforce crisis is catastrophic, or that it is indefinite. Across history, humans have always evolved in response to technological change. However, organizations should prepare for the transition between the current state and the "automation aftermath" to avoid a turbulent ride.

Given these statistics, now do you believe that the Acme scenario described at the beginning of this chapter could take place? How likely is it that growth in your industry or field will be stunted by a lack of qualified employees? How will the global skills shortage impact your organization?

The velocity of change in fields like robotics, data science, and digital security requires a highly skilled workforce. This raises important questions that leaders across the public and private sectors must address and collaborate on, such as:

- Who's responsible for re-skilling the workforce as millions of jobs are eliminated or radically changed by automation in every industry from health care, to manufacturing, to hospitality, to transportation?
- What rights and/or benefits should "gig economy" employees be entitled to?
- How can we make our labor, immigration, and human capital policies more flexible to accommodate the rapid changes impacting all industries?

These are big questions for which there are no easy answers. Particularly when these questions require resources for retraining and reeducating the workforce of the future.

At this point, I've laid out a case for workforce planning with the Acme case study and the data from these sobering labor reports. I've also introduced the argument that this type of high-velocity change will require more future-focused, agile leadership, particularly around workforce planning. To tie these two lines of thinking together, I want to turn now to examine some strategies that you can use to start practicing high-velocity leadership. The following section is intended to help you future-proof your own role as a leader, and to prepare your organization for the Great Restructuring of tasks and talent in the future of work.

Becoming an Agile, High-Velocity Leader

Let's drill into the concept of high-velocity leadership a bit more. As you recall, the three characteristics that define high-velocity leadership are speed, direction, and acceleration. Each of these elements poses unique demands for the work that leaders must do to be successful in a 4IR environment.

Speed is the first element of high-velocity leadership. It's common knowledge that the pace of change in organizations has been steadily increasing since the 1960s. The increase in pace corresponds with the adoption of computing and mobile technologies, and computing power doubling every two years, a phenomenon known as Moore's Law. In 1965, Gordon Moore observed that computing power was doubling every year. In 1975, Moore revised his prediction that processing power would continue to double every two years.

This trend continued until 2015 when the doubling of processing power slowed to every two and a half years.

To put this exponential growth in computing power into perspective, let's say you deposited a penny into your savings account on September 1. Then on September 2, you doubled your deposit and put two pennies into the same account. On day three you doubled again and put four pennies into your account. If you repeated this doubling of penny deposits into your account, after your deposit on September 30, you would have more than $10.74 million saved. Now, that's fast money! Computing power has grown at a similar pace, only instead of doubling every day, it doubles every couple of years. This extraordinary pace of growth has not been observed in any other facet of life or technological advancement before in human history. This is why fourth industrial mega trends are so much different than mechanical mega trends of previous industrial eras.

This speed of growth and processing power has led some industry experts to hypothesize what forces will continue to shape the global workforce. For example, research BCG conducted predicts that six forces will reshape organizations' demand for talent: (1) automation, (2) big data and analytics, (3) access to information and ideas, (4) simplicity in complexity, (5) agility and innovation, and (6) new customer strategies. Speed is the unifying factor among BCG's six critical forces. Each of these forces is a significant contributor to the wave of high-velocity change that leaders must learn to ride.

In support of the working hypothesis in this chapter, the following discoveries made by BCG in 2017 are very helpful:

- Ninety percent of managers say that agility is critical to strategy execution, indicating that a nimble organization can go faster (speed) than one that is rigid.
- By 2020, 7.6 billion people will use 11.6 billion mobile devices. This means that access to and speed of transactions across these devices will be faster than ever before. This will be compounded by the development of 5G mobile technology. For the first time in human history, the number of connected devices will exceed the number of people on the planet.
- And, finally, BCG reports that half of all jobs in the United States could be automated by 2050. These findings reinforce that speed is the single unifying force shaping the workforce: speed of automation and speed of hiring, training, and redeploying talent are monumental challenges for 4IR leaders.

What Are the Benefits of Speed?

Smart, connected leaders must be aware of the advantages and disadvantages of speed in driving high-velocity change. When leaders are aware of the risks and virtues of speed, they can use speed to their strategic advantage to become more agile and outrun their competition. Let's start with the advantages of speed.

First, speed creates stability for an object in motion. For example, think of riding a bicycle at a very slow pace. The bike is unstable, right? You feel like you are going to fall off the bike when you're going too slow. Therefore, leaders committed to building agile teams and organizations must assume that the pace of organizational life will at least remain a constant, if not increase. That is, the pace of organizational life isn't going to get slower. In fact, the pace of technological change is likely to increase rather than decrease. This means, as my colleague and change management expert Karl Schoemer always says, "the pace of change today is the slowest it will ever be in your career. Get used to it!"

The second benefit of speed is that it will force organizations to become more agile. As American management consultant W. Edwards Deming is often quoted as saying, "It's not necessary to change. Survival is not mandatory." Leaders interested in survival and in making agility a source of competitive advantage have to ensure that a few things happen. First, products must be designed faster and better. Second, learning from failure must happen faster. Third, human resources processes must get faster and better at finding and placing the right talent in the right roles. Finally, leaders must find better ways to support rapid learning across all levels of the organization to improve efficiency and effectiveness. Agility begets speed and enables products, services, and experiences to be redesigned and redeployed faster. Fast is the name of the game. And building and maintaining speed in a 4IR world requires agility.

The final benefit of speed is that it creates momentum. If agility begets speed, then speed begets momentum. Organizations that get stuck are those that fear speed and change. A few months ago, I was on-site with a leadership team, helping them develop future-proofing strategies around change leaders. My client pointed out that her organization's problem wasn't "the flavor of the month." She said their problem was that they had chosen "one flavor" and weren't interested in trying any others! This comment was telling about their culture and the lack of momentum for change they were able to create. No agility, no speed, no momentum. Organizations that move faster, by definition, must try new ways of working and new "flavors" to get the change flywheel in motion.

What Are the Risks of Too Much Speed?

When leaders say, "go faster," most followers say, "slow down!" This begs the question of how much speed is too much? The most agile leaders I know aren't speed or adrenaline junkies. They actually have a very keen ability to assess the risks of too much speed (or not enough speed), and they know how

Benefits

SPEED
Figure 3.1

to adjust the organization's pace. Here are some concepts and tools to help you improve controlling an organization's speed at scale.

If you think too much speed is risky, you're right. The relationship between leadership actions, speed, and the benefits is a curvilinear rather than linear relationship. I call this the "too much of a good thing phenomenon." Like drinking beer or eating chocolate chip cookies, a good thing is usually good up to a point, but then you hit a wall and the benefits start to diminish. For example, after three beers or glasses of wine you might be the life of the party and even become a good dancer. This "good thing" may have social benefits. However, after consuming four, five, or six drinks, the benefits of alcohol start to diminish, and you start introducing unnecessary social risks. You get the picture.

Using the same "too much of a good thing" relationship, Figure 3.1 depicts the relationship between the speed with which leaders drive change and the organizational benefits.

The secret to making speed an asset to your organization is captured in a tool that I call the 3Ms of Agile Leadership. The 3Ms are: **monitor**, **modify**, and **maintain**. These three leadership behaviors can help you optimize the speed of change, the speed of learning, and the speed of innovation in your organization.

Tool #8: The 3Ms of Agile Leadership

- **Monitor:** Agile leaders monitor the impact that speed is having on their followers. This is an offensive strategy for keeping your finger on the pulse of the heart of your organization—your people. Organizational agility surveys;

smart, connected dialogues; innovation round tables; and executive listening sessions are useful monitoring tactics for assessing the impact that speed and velocity are having on your people. Listen to people's opinions and use them as a speedometer.

- **Modify:** Agile leaders use data-driven methods to know when to modify the pace of change. Workplace analytics and econometrics can tell you when people have greater capacity for more speed. Some feedback warns us to slow down. Sometimes analytics tell us to change course. How do you know when you've exceeded the speed limit for your organization? Remember this rule of thumb: "Eight you're great, nine you're mine." This rule applies to exceeding the traffic speed limit before you get pulled over, and it applies to modifying the pace of change in your organization. In other words, if your employees are exceeding 80 percent of their total capacity for change, then you're going too fast. There isn't time for thinking, reflection, or maintenance. Monitor your organization's speed with the right kind of descriptive and predictive analytics.

- **Maintain:** Agile leaders invest in organizational maintenance of their people. At the risk of reducing complex human behavior to a set of mechanisms that can be "fixed," maintenance is about rebuilding, repairing, and renewing the behavioral ecosystem of your organization. If you're trying to squeeze every bit of energy out of your people, you WILL burn out your employees. Maintenance is about slowing down to reflect, repair, and recharge so that your employees don't suffer from adrenaline overload or burnout. Maintenance begins with identifying employees' basic workforce needs (e.g., clear behavioral expectations, rewards, recognition, and mutually beneficial learning and development opportunities). Once you know what people need, equipping team leads with the tools and resources to fulfill those needs is essential for maintaining employee resilience, well-being, and agility.

Mastering Direction in an Age of Uncertainty

Direction is the second element of high-velocity leadership. You can look at direction in two ways. Conventional thinking about direction brings to mind images of leaders who "give" direction or who direct other people's actions. This kind of direction is reminiscent of Frederick Taylor's notion of *management knows best* and workers are dumb as oxen. Although there will always be instances in which leaders must "take charge," 4IR leadership is less about directing others than in previous industrial eras.

The second way to think about direction in a 4IR world is as a line of movement from here to there. I like to think of direction in terms of the four cardinal directions: north, south, east, and west. In this sense, direction is a matter of degrees. Shifting your organization's direction one or two degrees will lead you to a very different destination.

This means that even the slightest shifts in direction each day have the power to unleash unforeseen opportunities and challenges. As such, leaders must ensure that they have strong navigational instincts and tools. When the destination is unclear, leaders must always fall back on their true north (i.e., the organization's purpose and values). When the line of sight along the path is foggy and unclear, leaders must learn to trust their instruments (i.e., your compass, GPS, analytics, and/or instincts). These tools help maintain momentum through uncertainty and provide a sense of stability and direction for those who follow your leadership.

The people who are following you have a deep need for direction. When teams of followers have a sense of direction, they tend to feel more energized and have a greater sense of stability and hope about the future. Leaders must learn how to build confidence among followers and communicate to them that the organization is traveling in the right direction. Building confidence in your organization's direction can be achieved by involving followers in setting a shared direction or vision, listening to customers' needs and discussing insights together, and using advanced analytics to predict where markets are headed. If your followers and employees do not feel listened to or engaged in setting direction, then strategy implementation and execution will suffer. Lack of involvement or lack of direction lead to low performance.

Today, no single leader or leadership team can operate with the confidence that they "know it all" or "can do it all." The "lone hero" leader is dead. Where the lone hero once stood is an empty space waiting to be filled by a more collaborative, curious, human-centered, smart, connected leader. Are you up for the job?

Smart, connected leaders:

- understand the power and limitations of their mindset about the future;
- work at being more fully present;
- know the basics of high-velocity change (speed, direction, and acceleration);
- learn from failure and take smart risks;
- involve followers in exploring and setting direction; and
- invest in their own learning and know how to develop their followers.

The smartest and most connected leaders invite critique and radical openness from their team. They have a well-developed self-concept and self-esteem, which allow them to be wrong while leading their team forward with humility and confidence. Smart, connected leaders create innovation meritocracies where the best ideas win—and the best ideas needn't be their own. Finally, smart, connected leaders know how to win the minds and hearts of their followers.

Winning followership when there is great uncertainty starts with setting direction. Leaders who want to become more agile, smarter, and more connected to their followers need to master the following to set direction amid uncertainty:

1. Master *listening* at scale. Listen to everyone: peers, customers, employees, competitors, collaborators, business-to-business (B2B) partners, and so on. Ask the right questions and know when, where, how, and with whom to maximize organizational listening tools like town halls and surveys to become more agile to change.

2. Master *shared visioning* and setting direction. Identify who your most future-focused employees are and maximize the mindset they bring to future-forecasting conversations. Who are your futurists, innovators, opportunists, observers, and historians?[9] What do they think the future holds? Involve critical stakeholders to determine the right direction and create excitement for that envisioned destination.

3. Master *acceleration* and deceleration. Once shared vision has been created, high-velocity leaders live by this acceleration rule of thumb: "brakes to slow" and "gears to go." They know what the brakes are in their organization and they know how to use them. Similarly, high-velocity leaders know how to generate organizational "torque" at low speeds, and how to skip gears to get up to maximum speed fast when a situation demands.

Now that we've tackled speed and direction, let's tackle acceleration, the third demand of high-velocity leadership.

The Paradox of Acceleration

The third variable of high-velocity leadership is acceleration. Acceleration is the rate at which velocity changes. It's nonuniform velocity. Think of riding a bicycle down a hill. As you roll down the hill, acceleration increases your velocity.

The paradox of acceleration is that it really involves two different sets of skills and two different mindsets about high-velocity change. The first mindset values speed and acceleration. This makes sense, as the word *agility* implies speed, quickness, and acceleration. The "faster-is-better" mindset goes something like this:

> IF WE ARE GOING TO COMPETE IN A FAST WORLD, THEN WE MUST BECOME FASTER AND MORE AGILE.

This logic favors quickness, nimbleness, high speeds, and acceleration. Without question, this is the most common mindset I observe among leaders. In the style of the old Aesop's fable between the tortoise and the hare, the faster-is-better mindset is that of the hare. This mindset is particularly pervasive in the business world—regardless of industry. It has driven industry executives to invest inordinate amounts of energy, time, and money into organizational design and restructuring initiatives to increase the speed and pace of acceleration with which they can execute their strategies.

Interestingly, as many leaders are discovering, not every employee shares the same mindset about the need for speed. Some leaders and individual contributors have more of a slow-and-steady, tortoise-like mindset. Even the most agile strategy and structure, with carefully designed roles and responsibilities, decision rights, and reporting relationships, will only take an organization so far. True agility and acceleration require a change in mindset, skills, and behaviors.

The change from a "slow-and-steady" organization to a "faster-is-better" agile organization, requires an understanding of the human side of change and acceleration. Put differently, acceleration requires that employees feel involved in and energized about the purpose and direction that their organization is traveling in. "Sprinting" requires excitement and belief in the outcomes a team is working toward. It requires building teams around the mindsets and skill sets of each individual member—that is, not everyone on the track team is a sprinter! Some people are long-distance runners, and some people are better at pole-vaulting. A change from slow and steady to fast and ready requires that employees believe in the design of their workflow. Employees must also feel they have the proper skills to be successful in the new way of working, and that they have a fulfilling career path within the new agile, high-speed, rapid-acceleration environment. All of this change, as it turns out, is very difficult to execute well.

And to complicate matters more, the paradox of acceleration is that at times, deceleration is actually a better organizational strategy than acceleration. The "pro-slow" logic goes something like this:

IF WE ARE GOING TO SUCCESSFULLY NAVIGATE COMPLEXITY, THEN WE MUST GET GOOD AT SLOWING DOWN SO THAT WE CAN GO FAST IN THE RIGHT DIRECTION.

Complexity is a 4IR challenge that requires skills around learning, discussing, sense making, discerning, and decision making. Going against the herd and slowing down can be unpopular at best and dangerous in the worst of cases. However, we needn't look very deep into history to see countless cases of where *groupthink* and a need to go fast have led to organizational failure.

Groupthink is a term that found its way into social psychological and team dynamics research primarily through the research of Yale professor Irving Janis. In a 1971 article published in *Psychology Today*, Janis defined groupthink as, "the mode of thinking that persons engage in when concurrence-seeking becomes so dominant in a cohesive in-group that it tends to override realistic appraisal of alternative courses of action" (p. 43). According to Janis, groupthink occurs when the members of decision-making groups refrain from harsh judgment of their leaders' or peers' ideas. Instead, they prefer a comforting "we" atmosphere.[10]

One of the most memorable illustrations of groupthink I can recall comes from the 1957 film *12 Angry Men*, starring Henry Fonda. If you recall the story, an 18-year-old man was on trial, accused of murdering his father. Early in the jury's deliberation, 11 of 12 jurors quickly return a guilty vote, which would have sentenced the boy to death. However, Juror #8, played by Henry Fonda, takes an unpopular stance of slowing down and asking his peers to reconsider the evidence before unjustly sentencing the young man to death. This deliberation and consideration of alternative actions is one of the defining moments in cinematic history, and a great illustration of the powerful force that groupthink can have on important decision making.

Outside of popular media, Janis's research explored the impact of groupthink on a number of real-life historical decisions made by groups. These included the Japanese attack on Pearl Harbor, the Bay of Pigs fiasco, the Vietnam War, and the 1986 NASA *Challenger* disaster. Each of these case studies illustrates how bad decisions were made because leaders did not slow down to consider alternative viewpoints prior to proceeding.

What Janis's work teaches us about groupthink is that the social norm of "concurrence seeking" or consensus building can lead to premature decisions. When I ask clients to describe their organization's culture and they say, "We have a consensus culture," my groupthink radar goes off. In addition to culture and consensus building, groupthink can also occur because complexity is underestimated. In other cases, simply failing to ask "what if" questions, engaging the ideas of out-group members, or considering the possibility that prevailing thought could be wrong can all lead to groupthink and to bad decisions.

The paradox, of course, is that deliberation and discernment take time and require slowing down. This can cause tension between "slow and steady" versus "agile and ready" leaders. This paradox raises the question: Does organizational agility only require the ability to sprint (i.e., a method common to agile software development), or does it also require the ability to "marathon" for the long haul? Or both? From what I've observed, the most agile organizations know when to use their breaks to slow and their gears to go.

The challenge of managing the paradox of acceleration/deceleration is knowing when to speed up or slow down. I will offer a number of strategies for managing this tension during decision making in the chapter on discernment. But in a global sense, the first step toward smart, connected acceleration is learning how to appreciate the value of both the "agile-and-ready" and "slow-and-steady" mindsets. Because the "faster-is-better" mindset is so prevalent in organizational cultures, it's worth listing a few of the benefits of deceleration here.

Deceleration has the following benefits:

- Risk mitigation
- Improved problem solving

- Opportunities for learning and development
- Team and stakeholder alignment
- Employee engagement and well-being
- Strategic thinking
- Better decision making
- More inclusive conversations
- Strengthened relationships and trust

Given these benefits, what tools can leaders use to slow their organizations' velocity through deceleration? In the previous section, I offered some valuable tools (e.g., the TFA loop, the P2PDG, Open Space meetings, etc.) scalable to teams and organizations that will help with deceleration. If you find yourself asking, Why are we struggling to execute this awesome strategy? Why is our decision process or decision quality not as good as it should be? Why are we making mistakes that are negatively impacting our customers or our brand? Why don't we have better collaboration in our organization?, then try using some of the tools from the chapter on presence to decelerate your team or organization. By being present with your peers and team and considering the evidence before you—as Juror #8, although unpopular, suggested—it might help your organization go slow to go fast—or simply avoid a crash.

At the end of the day, smart, connected leaders are good at both organizational acceleration and deceleration. These elements of high-velocity leadership are critical for managing the speed of change in an organization. Remember that the relationship between speed and its relative benefits is curvilinear: speed, like all good things in life, must be approached in moderation. If you want to get to where you're going in one piece, then you have to know when to speed up and when to slow down.

Connecting Speed, Direction, and Acceleration

So far in this section, we've focused on three elements of high-velocity leadership: speed, direction, and acceleration. Mastering these three elements of high-velocity leadership demands that leaders develop a mindset around the benefits of "agile and ready" versus "slow and steady."

Figure 3.2 provides a simple depiction of the outcomes that leaders will achieve when all three elements of high-velocity leadership are in place versus when one element is deficient in the high-velocity leadership equation. This equation might look familiar to leaders who have studied change management. Similar models have been used around the elements of successful change management (e.g., vision, stakeholder inclusion, etc.).

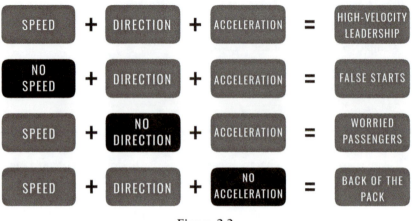

Figure 3.2

Tool #9: The High-Velocity Leadership Equation

The High-Velocity Leadership Equation is a powerful tool for communicating the value of the three elements of high-velocity leadership, and explaining why speed, direction, and acceleration/deceleration are so critical for future-ready organizations and leaders.

When all three elements of high-velocity leadership are in place and appropriately leveraged, leaders can optimize speed, clarify direction, and leverage acceleration/deceleration to improve performance and results. However, when leaders fail to build traction and speed, even when they feel they have a clear direction and their foot on the accelerator, they may experience false starts or no movement at all. They're simply revving their engines with the hopes of going somewhere. Gaining traction and building momentum are essential first steps of high-velocity leadership.

On the contrary, when leaders have lots of acceleration and speed, but no direction, they're going to have a lot of worried passengers on their bus—regardless of whether those passengers are "in the right seat." Acceleration plus speed without direction creates a sense of instability among followers. This formula will likely create motion sickness, active disengagement, and turnover among followers. The consequences of "fast-and-furious" leadership that lacks direction are costly to your organization. Hence, there is value in knowing when to slow down to check the road map. Just as in driving a car, you should never look down to check your GPS or map when you're doing 80 miles an hour. Slow down, correct course, then accelerate.

Finally, if a leader sets a shared direction and has the right speed but is unable to push on the accelerator to pass the competition, the organization will likely fall behind in the race. Without acceleration, which requires energy and the proper timing, organizations cannot pass their competition or lead the pack in the eyes of their customers or constituents.

Measuring Your High-Velocity Capabilities

What are your strengths and opportunities relative to high-velocity leadership? Take the High-Velocity Leadership Self-Assessment to find out. This tool measures every aspect of high-velocity leadership that is important. Use the scoring guide to compare yourself to your peers, or incorporate them into your next "360" feedback assessment.

Table 3.1   Tool #10: High-Velocity Leadership Self-Assessment

Measure	Item	Scale
Priorities	My priorities have been clear during the past 30 days.	1 2 3 4 5 Strongly Disagree Strongly Agree
Direction	I am confident in the strategic direction that my organization is pursuing.	1 2 3 4 5 Strongly Disagree Strongly Agree
Involvement	I feel actively involved in decisions that impact my work.	1 2 3 4 5 Strongly Disagree Strongly Agree
Monitor	There is excellent communication at all levels in my organization.	1 2 3 4 5 Strongly Disagree Strongly Agree
Modify	Our organization is better at changing direction than other organizations I've worked in.	1 2 3 4 5 Strongly Disagree Strongly Agree
Maintain	In the last 30 days, I have had structured time to reflect and recharge my energy levels.	1 2 3 4 5 Strongly Disagree Strongly Agree
Speed	The speed with which our organization is making changes is perfect for me.	1 2 3 4 5 Strongly Disagree Strongly Agree
Acceleration	Our leaders have good instincts for speeding up or slowing down at the right times.	1 2 3 4 5 Strongly Disagree Strongly Agree

Scoring guide

1.   Add up your total score for the eight items (total of 40 points possible).

2.   Use the following as a guide for interpreting your score:

   •   8 to 16 = Your organization needs help preparing for high-velocity change.

   •   16 to 24 = You and/or your organization have some significant areas for improvement.

   •   24 to 32 = Not bad. Still plenty of room for growth.

   •   32 to 40 = Congratulations! You're a high-velocity leader in a high-velocity organization.

3.   Now share your results with others and discuss what you can do to ramp up your leadership for the 4IR, and prepare your followers for high-velocity change.

Preparing Followers for High-Velocity Change

Now let's take a look at what you can do to prepare followers in your organization for the type of high-velocity journey that lies ahead. First, let me say a bit more about my use of the term *followers.*

Followers in a 4IR organization are influencers, innovators, creators, and doers in your organization. You have to let go of the mindset that "follower" means a passive spectator. This is no longer the case. Followers are the new leaders. Followers in the digital age are active agents that are necessary for great leader–follower partnerships. The types of followers you want to win are the people who motivate others, who want to be a part of your organization's success, and who have extraordinary ideas and special skills. You need a critical mass of these types of followers in your organization to feel, think, and behave in this way. This requires collaboration (the theme for the next chapter) and a renewed mindset about shared power between leaders and followers.

Based on my consulting experience, world-class organizations have at least 60 percent of their employee population who think, feel, and act like smart, connected followers. This is the critical mass needed for creating an organizational tipping point. The real benefit of having 60 percent of your organization who lead, influence, and engage others around them is that they are more ready, willing, and able to respond to high-velocity change. They're more resilient and more likely to drive change forward, rather than merely "going along" with change.

I tested this "resilience" hypothesis with a client recently. The organization was a global company with about 90,000 employees. Although only about 22 percent of its workforce consisted of highly engaged followers, these employees were **three times more likely** than actively disengaged followers to report that they were "doing things to actively drive the change forward." I don't know about you, but I'll take 3x odds on a successful M&A, digital transformation, or reorganization any day of the week! What's more: through this experiment, we were able to statistically identify the specific factors that inspired engaged followers to actively drive change forward. The results might surprise you. These employees weren't motivated by monetary bonuses or free dry cleaning. They were motivated by a clear purpose and involvement with setting the direction of change in their organization. This case proves that investing in your followers and meeting their needs is critical to driving high-velocity change.

My message to this organization's leadership team was simple: "Think how much faster and farther you could move this organization if you tripled the number of leaders, influencers, innovators, and doers in your company. Think of how much energy for change you are leaving on the table!" The findings from our study allowed the executive leadership team to make some simple management shifts and modest investments in team effectiveness and change leadership education to get a critical mass of their followers ready for high-velocity change.

A How-To Guide for Getting Followers Ready for High-Velocity Change

My research, like the previously discussed study, has shown me time and again that the most effective organizations are driven by a compelling purpose. A clear purpose ensures that all leaders, followers, and customers are clear about what the organization stands for, and the direction that it is going in. This clarity of purpose attracts would-be followers and customers.

I also believe that the best organizations are clear about who their customers are and they have made a pledge to meet their customers' needs. And, finally, I believe that the best organizations are filled with leaders and followers who create cultures that foster energy, inclusion, and excitement about the work that the organization is doing. If you can do these three things, you are ahead of the majority of leaders and organizations in the world.

The first step in preparing others for high-velocity change is to get your purpose, promise, and culture right. This step is critical. Don't underestimate it. Every year companies spend billions of dollars on leadership training to "develop and retain" their best and brightest leaders, and they end up losing those leaders to competitors. Why do organizations train leaders for the competition? Because the company didn't have a compelling enough purpose, promise, or culture to invite those superstars to stay.

Your purpose should be simple, clear, and compelling. It should answer two simple questions:

- Why does your organization exist?
- What does your organization do to make the world a better place?

You might be saying, "Well, we have a mission, vision, and values, so we can tick the box on 'purpose.'" Wrong! Your purpose is *why you exist*, your mission is *what you hope to accomplish*. The difference between them is huge in winning followership and future-proofing your workforce against high-velocity change.

Let me break down the four concepts of mission, vision, values, and purpose for you. The following highlights the key differentiators between them:

- Mission describes **what** your organization is and what it hopes to accomplish.
- Vision describes what your organization continually strives to **become in the future**.
- Values describe what your organization **cares about** and puts a premium on.
- Purpose is a simple statement of **why your organization exists**.

Start preparing your organization for high-velocity change with a clear statement of why your organization exists. If you don't know why your organization exists, how do you expect your followers or customers to know?

Purpose is the most compelling reason to get people to join your organization and to work hard with you to achieve your mission. It's a compelling reason to get customers to buy from you and constituents to believe in you. Purpose is essential for attracting the best people and customers to help you succeed in a 4IR world.

If you need ideas about how to find a compelling purpose for your organization, plenty of resources on this topic are available to you. For example, Simon Sinek has communicated a simple and compelling message about the power of purpose. His TED Talk based on his book *Start with Why* is among the most viewed videos on YouTube. Watch his video and start defining "why" your company exists and use it to inspire and win followers.

The second step in preparing your team members for high-velocity change is to define your promise to followers—whether they are customers, constituents, or key partners. This should be informed by and aligned with your purpose. Your promise should also be something that can be clearly described. What is the experience you promise to deliver to employees or customers? What is the quality of the product or service you will deliver? A clear promise of value helps people inside and outside of your organization hold leaders accountable for doing the right things.

The third and final step is to create a **behavioral ecosystem** or **culture** that is aligned with your purpose and promise. A strong culture isn't just a nice complement to your strategy; it is how work gets done and how strategies are carried out. Culture simply describes the behaviors that are *expected* and *accepted* within a social ecosystem. It's that simple. No need to complicate it or to pay millions of dollars to consultants to help you build an ideal culture based on their benchmarks or pseudoscience. The best culture is the one that leaders *expect* based on their purpose and promise to customers and employees. And, if this is the culture you decide you want to create, then you cannot *accept* or tolerate behavior that works against your purpose or promise. It's that simple. You get what you except and accept.

Defining your culture in these terms helps you create and manage clear expectations about "how we roll" as an organization. Your culture should be able to be spelled out in a short list of behavioral expectations. Some clients call these "leadership expectations," "guiding behaviors," "leadership behaviors," or "values in action." Do your leaders and employees know what behaviors are expected of them? Are they living these behaviors every day? Are these behaviors being taught and developed through the various touchpoints in employees' experience? Are leaders and employees being held accountable to this behavioral standard?

Netflix has a very nice example of how they've set these simple behavioral expectations that align with their core values. You can link to Netflix's behavioral expectations here on Slide Share: www.slideshare.net/reed2001/culture -1798664.

As you'll see in the presentation, Netflix clearly states that it values "judgment." That's great. I don't know too many leaders who would say they *don't* value good judgment. But what does judgment look like in the context of Netflix's culture and promise to its users? As stated in the Netflix presentation, judgment means, "You identify root causes and get beyond treating symptoms"; "You smartly separate what must be done now, and what can be improved later." These behavioral expectations "show rather than tell" everyone in the company what will be *expected* and *accepted.* Can you describe your organizational values with this degree of specificity, clarity, and precision? Can the average employee in your organization do the same? Are your value and behavioral expectations future-ready? Are your values designed to be put into action?

Clear behavioral expectations are essential in preparing your workforce for high-velocity change. Clear behavioral expectations create several benefits for your organization:

1. They're observable—you can describe where you see your values being lived out and where you don't.

2. They're teachable—you can coach leaders, managers, and individual contributors around the behaviors that you value and those that will enable greater organizational agility.

3. They're measurable—you can hold followers accountable for them and recognize them for getting the right work done and in ways that are consistent with your expectations.

4. They're marketable—you can attract people to your organization by sharing your story and showing people what your company stands for—your purpose.

5. They're stable—you and your employees can rely on the fact that, even as the world changes, these behavioral expectations will remain relatively stable.

Most people don't think about the hidden value in #4 (They're marketable). But, as Netflix's simple "culture deck" proves, describing your culture can get you big attention in the war for talent. More than 16,000,000 people have viewed and learned what Netflix stands for on Slide Share alone! That's a lot of free press. This makes Netflix a magnet for attracting the best talent in the world. Well, that and they make really cool original series.

In Table 3.2, I've included an organizational assessment, a checklist of sorts, to help you get your organization ready for high-velocity change. Use this to help you and your leadership team determine whether your organization's purpose, promise, and culture are clear, compelling, and consistently communicated to your employees and customers. This assessment can also help you to determine whether or not your followers are ready for high-velocity change.

Table 3.2   Tool #11: High-Velocity Organizational Readiness Assessment

	ITEMS	YES	NO
PURPOSE	Do we have a clear purpose?		
	Is our purpose truly distinct from our competition?		
	Is our purpose distinct from our mission, vision, and values?		
PROMISE	Do all employees understand what we want to deliver to our customers?		
	Are our employees actively advocating for our brand with customers?		
	Do we have measures to determine how consistently we're delivering on our promise?		
BEHAVIORAL ECOSYSTEM	Do employees have a clearly stated set of behavioral expectations that they will be held accountable for?		
	Are these behaviors aligned with what we say our organization's purpose is?		

Table 3.2   Tool #11: High-Velocity Organizational Readiness Assessment

ITEMS	YES	NO
Do we have education in place to teach these behaviors to our community of followers throughout their careers?		
Do we have measures in place to hold everyone in the community accountable to these standards?		

If you answered "No" to any of these items, then you probably have some work to do in future-proofing your organization and preparing for high-velocity change.

This section has covered how to get your organization ready for high-velocity change. Purpose, promise, and culture are the most important elements for preparing the organization. Of course, there are other elements for future-proofing your organization against high-velocity change. These will be explored in subsequent chapters. Because this is a leadership book, I've started with purpose, promise, and culture because these are elements of change readiness that only leaders and boards can control. These are the most strategic levers that require smart, connected leadership for building the organization of the future.

Chapter Summary

1. There are three important elements of high-velocity leadership: speed, velocity, and direction. Smart, connected leaders must master all three of these elements for winning followership and driving change. This chapter presents useful statistics about the speed with which organizations are changing due to 4IR mega trends.

2. Speed has benefits and risks. There is a curvilinear relationship between speed and its benefits. High-velocity leaders know how to accelerate and decelerate their organization to control their velocity and direction. Controlling speed creates stability for followers.

3. This chapter introduced four new tools: The 3Ms of Agile Leadership; the High-Velocity Leadership Equation; the High-Velocity Leadership Self-Assessment; and the High-Velocity Organizational Readiness Assessment. Use these tools to prepare yourself, your team, and your organization for the demands of high-velocity leadership and change.

Recommended Actions

1. Discuss the statistics in this chapter with your HR leaders and your leadership team. Learn the roles most vulnerable to disruption in your organization/industry. Learn how the speed of the 4IR will impact your workforce.

2. Start conversations about whether your organization has a "slow-and-steady" or an "agile-and-ready" mindset, and how that mindset might be serving your organizational strategy.

3. Share the High-Velocity Leadership Equation with your team and discuss where your team is struggling or strongest.

4. Invite your leadership team to take the High-Velocity Organizational Readiness Self-Assessment and have a conversation about what the results mean and what to do about them.

5. Schedule an off-site meeting with your leadership team to check the alignment between your purpose, promise, and behavioral ecosystem. If you don't have a clear purpose, promise, culture, or behavioral expectations for delivering on these, then have an outside expert conduct a readiness assessment for you. I'm happy to recommend some of the best in the business.

CHAPTER FOUR

Collaboration

Dimitri is a VP of sales for a medical technology company. As the head of sales for one of the highest revenue-producing divisions, he leads a group of sales managers and collaborates with marketing, engineering, and research and development. For the last 15 years, Dimitri's division has had complete autonomy in running all facets of their part of the business. They hired their own people and could pull all the levers that impact the profit and loss (P&L) statement without having to ask for permission. With this freedom, they controlled product development, marketing, sales, and distribution to customers. But six months ago, Dimitri's world changed. The company restructured and launched a new global operating model.

In the new operating model, Dimitri works in a highly matrixed organizational structure, reporting to the unit president and a global head of sales strategy. Similarly, the heads of finance, marketing, distribution, engineering, and research and development (R&D) all have a matrixed reporting relationship to their respective enterprise leader and the unit president. As Dimitri and his peers describe it, the new organizational structure is "messy" and they're still "working the kinks out."

The "messiness" rears its head in the form of unclear decision rights, conflict, and tense conversations about priorities among leaders and their teams. People spend a lot of time and energy in meetings complaining about the structure instead of carrying on with the work of building the business.

In certain parts of the organization, "turf wars" between different business units have started to emerge. Leaders within these business units feel like they are competing for enterprise resources so that they can satisfy their customers' needs. Ironically, instead of breaking down the "silos" (barriers between divisions) in the organization and speeding up decision making, the new matrix structure seems to have had the opposite effect. Information is being hoarded between the divisions, decisions take longer than ever, and leaders between the different divisions seem to be more entrenched than ever before.

Dimitri feels like he has lost his control and power to the corporate leaders who aren't as close to his customers or to his sales team. And his sales managers are disengaging because they feel like their hands are tied, and that they can do very little to keep their sales reps motivated. The sales reps are frustrated because they don't feel like the business unit is innovating fast enough to keep up with the competition in their product lines. Dimitri is also getting feedback that many of his best sales reps are being recruited by competitors who are promising bigger payouts based on their competitive product line advancements. Dimitri finds himself getting more and more frustrated with leading in this environment.

The challenging experience of leading in a matrixed environment is more common today than ever before in the history of large-scale organization. In fact, a Deloitte survey of more than 7,000 executives in 130 countries found that more than 80 percent of respondents were either currently restructuring or had recently finished the restructuring process.[1] This same study found that 70 percent of those reorganizations fall short of delivering their promised value because of executive disengagement, disobedience, and frustration. In other words, organizations fail at creating efficiency, effectiveness, and collaborative advantage because they get the people side of cross-functional collaboration wrong.

This section covers the topic of why collaboration is so critical to future-ready organizations. *Collaboration involves a process of coordinating work to achieve shared goals, shared value, and shared futures.* These are ever-increasing demands for leading in the fourth industrial revolution. This section specifically addresses cross-functional collaboration between teams within an organization.

Shared goals, value, and futures are important aspects of collaboration in the fourth industrial revolution. Teams work together more effectively in achieving shared goals, for instance, implementing a new enterprise resource system, when they have shared value and an implied shared future between collaborating parties. However, when collaborators don't see shared value and/ or a connection in the future, then the results of the collaboration suffer. It's leadership's role to ensure that cross-functional collaboration is, first, justified. That's right, not all collaboration is necessary. When collaboration isn't warranted, leaders should discourage it. However, when collaboration is necessary, then leadership must communicate both the shared objectives and the shared value that it will create. Finally, leaders must communicate shared value in such a way as to inspire belief in, hope for, and clarity around the shared future resulting from the collaboration. Most leaders fail to effectively communicate shared value and shared futures when attempting to coordinate effective cross-functional collaboration. This is why most collaborations fail to generate the intended value and outcomes that they should.

To illustrate my point, let's look at an example of implementing a new enterprise resource planning (ERP) system in Dimitri's organization. This is a common example of collaboration across many organizations as they become increasingly digital. The first step is to establish a **shared goal**. The shared goal is relatively clear: work together to implement the system on time and on budget for the good of the organization. What makes a goal "shared" is that everyone understands the goal, and they feel they have some skin in the game, that is, that they will be rewarded for achieving it. Making this goal "shared" requires effective communication from leadership about the shared objectives of the system. This means building basic understanding by using lots of informative communication about what's changing, why, and building a case for that change. But effective communication about shared goals isn't simply one-way. It also requires seeking feedback, input, and giving all stakeholders a chance to have their voices heard.

Success will also require leaders like Dimitri to build a sense of **shared value** among all stakeholders. For example, the people in finance and the people in IT might not agree as to the shared value that the new system will create. The system might be seen as creating more value for the finance team than for the IT team, and as such the IT team might not commit wholeheartedly to the collaboration due to a perceived lack of shared value. Similarly, the dialogue regarding shared value will have to be coordinated and consistently communicated by stakeholders in finance, IT, HR, and business leaders like Dimitri. And, as with the informative communication about shared goals, the communication about shared value must also involve two-way dialogue.

Finally, to build the maximum level of engagement and commitment to implementing the ERP, leaders will have to foster a sense of **shared future** among all the necessary collaborative partners. Shared futures are commitments between two or more parties. Shared futures say, "We are in this together." Shared futures are mutually beneficial, which is to say they contain shared value. And to motivate and inspire all stakeholders, shared futures must be communicated, discussed, debated, aligned with shared values and goals, and contain a common purpose.

Without shared value and shared future, people really don't care much about collaboration. If *consensus* is good, and *collaboration* is better, shared futures address a deeper sense of *community* that exists beyond collaboration. As with all systems implementations, the work is never done. The changes never stop. Therefore, building an organization of the future requires full engagement to implement new ways of working, like implementing an ERP system quickly so an organization can tackle the next challenges, all in the name of a better shared future (for employees, customers, and the world).

Given the importance of shared goals, value, and futures, this chapter will go deep into the mechanics of cross-functional collaboration. We will not go into collaborations between different organizations or between public and private-sector organizations. Cross-organization and cross-sector collaborations are extremely important types of collaborations that will be essential for creating future-ready collaborations, communities, countries, and the world. Indeed, a true sense of "community," be it on a local or global scale, is only possible through top-notch interorganizational collaboration. But cross-organization collaboration is only possible once the fundamentals of collaboration have been mastered. Addressing these more complex forms of collaboration is simply beyond the scope of this chapter. This topic alone could fill the pages of an entire book!

My rationale for focusing exclusively on cross-functional collaboration is because I believe that leaders must first learn to collaborate within their own teams and organization before they can effectively collaborate with leaders from other organizations. From what I've seen, leaders have a hard enough time working together to achieve their own organization's strategic objectives and fulfill their own organizational mission. Collaborating across boundaries, sectors, and organizational interests is wishful thinking until all can play nice with classmates in their own school. In helping teams develop collaboration competence, I've also learned that we have to start somewhere. So, we might as well start with ourselves and those around us.

Once leaders in your organization have built collaboration competence and a collaborative mindset, then they can move on to building collaborative partnerships, teams, cross-functional collaborations, cross-business collaborations, and so on. In other words, leaders have to first focus on the fundamentals of collaboration before getting into advanced collaboration work.

Given this, my hypothesis is that if we (that is, me and people like you who are committed to helping leaders create future-ready organizations) start by helping individual leaders and teams, then we will support more efficient and effective organizational collaboration. By devoting an entire chapter to collaboration, I want to stress the importance of it as a key enabler of deeper forms of connection and community. I am confident that, with this awareness and foundational skill, more complex cross-sector collaborations will go more smoothly. More research, experience, and practical advice are needed on the mechanics of cross-organization and cross-sector collaboration.

In the remainder of this section, I will explain why collaborative advantage is an essential building block for future-ready organizations, and, as in previous sections, provide you with tools and tactics for creating collaborative advantage in the fourth industrial revolution. Along the way I'll debunk some of the most common collaboration myths and teach you how to overcome common collaboration barriers.

The Case for Collaborative Advantage

Earlier I provided a simple definition of *collaboration* as a process of coordinating work to achieve shared goals, create shared value, and build shared futures. *Collaborative advantage* is a response to the business strategy of creating "competitive advantage" within a market or industry. Global corporate strategy expert Todd Zenger, who is chair of Strategy and Strategic Leadership at the University of Utah's Eccles School of Business, has written a fine book called, *Beyond Competitive Advantage: How to Solve the Puzzle of Sustaining Growth while Creating Value.*[2] In his book, Zenger argues that traditional "positioning" strategies, like those of Walmart and Southwest Airlines, can straitjacket organizations, making it harder for them to sustain value creation over time. The goal of organizations, Zenger argues, should be to sustain growth, not by competitively holding onto one's market position, but by continuing to create value.

Establishing *collaborative advantage* is one such strategy for creating and sustaining value. It requires forward thinking, piloting, coordinating, prudent risk taking, and creativity. Creating collaborative advantage harnesses the power of relationships, connections, new technologies, and new possibilities to reposition, deposition, or completely launch an organization into an entirely new market or industry, thus creating new forms of value and sustaining the organization and people who contribute to it.

The value that collaborative advantage creates comes in several different forms. Perhaps the most obvious reason to foster collaboration, as implied thus far, is better operations. When information is shared more efficiently and decisions can be made more effectively, operations improve as does the experience for customers, constituents, students, patients, and so on. Better operations are the result of better knowledge management, more efficient and automated access to data and analytics, faster decision making, better prioritization, and improved customer experience.

Collaborative advantage also results in improved problem solving and innovation. When done well, collaboration empowers different teams to maximize their best and most creative thinking. Many companies are looking for "smart creatives" or "social quants" to drive innovation and imagine new possibilities. Many people have a misperception of innovation coming miraculously from the brain of a singular creative genius. Generations have celebrated and mystified the creativity and innovative genius of people like Thomas Edison and Albert Einstein, and contemporaries like Steve Jobs, Sir Richard Branson, and Elon Musk. But anyone can come up with a concept or an idea. Heck, you can steal or borrow a concept from a competitor. The real advantage and value of an idea comes from turning it into something that other people can use. Collaborative advantage, in the form of innovation, comes from turning a vision into reality. This takes coordination, communication, persistence, and lots of help from people with diverse talents and skills. Innovation is a team sport.

Finally, collaborative advantage is characterized by improved productivity and output. Collaboration strategies, structures, skills development, and new workflows and processes are only valuable if they produce better outcomes.

Collaboration simply isn't worth the effort if it doesn't make people, teams, and organizations more productive. Productivity can be measured in different ways in different kinds of organizations. In some organizations, like Dimitri's medical technology business, productivity might be measured in new sales. Productivity could also be measured in new ideas, products, and/or a better customer experience. When taking the first step toward designing an organization for collaborative advantage, leaders must ask themselves, will this improve productivity as we measure/assess it? If the answer is "no," then seek another strategic organization design because collaboration requires too much effort if it doesn't produce value in the form of increased productivity.

Creating collaborative advantage requires designing your organization in new ways (i.e., new strategies, structures, people, workflows, rewards, etc.). As Dimitri's case shows, organizations around the world are attempting to create collaborative advantage by designing more efficient and effective organizational structures. The research I've done for this book has taught me that leaders are making some pretty important discoveries about the challenges of creating collaborative advantage through restructuring. These discovering include the following:

1. Organizations are getting flatter, leaner, and more dynamic in hopes of creating collaborative advantage through efficiency, effectiveness, and local decision making. But leaders are discovering that structure does not necessarily equate to speed and success.

2. Matrix structures are more efficient and effective, in theory, but more difficult to lead in due to loss of power, authority, and control.

3. Silos and decentralized decision making can lead to poor performance, but so too can centralized decision-making structures. There are no guarantees or perfect models of decision rights so long as human beings are involved.

4. New structures that are flatter and more integrated require a different kind of leadership behavior, communication, and a different leadership mindset.

5. In the future of work, leaders must learn to master the matrix by influencing through relationships and ideas rather than through control and position.

These discoveries illustrate that "efficient and effective" structures, although a necessary element of a future-proofing strategy, are insufficient for creating the type of collaborative advantage that organizations need for success in the future. Deep collaboration requires a very different mindset, leadership behaviors, and ways of working in teams.

The most important part of using collaboration to create a future-ready organization is to build capability among people around communication, teamwork, shared power, and commitment management. These fundamentals are critical for achieving shared results, creating shared value, and building shared futures. Leaders must learn to foster an environment where shared value is the goal of every collaborator, and one where power relations are understood and effectively negotiated. This is the best way to optimize future-focused strategies to create collaborative advantage, as well as to create more efficient and effective organizations. However, mastering power relations and collaboration can be a challenge for many leaders because hierarchy is all they've known throughout their entire careers. When all you've experienced is command and control, it's hard to imagine what communicate and collaborate look like in practice.

Many executives assume that the next generation of leaders (i.e., the Gen-Xers and millennials) will simply adapt well to collaborative structures and become more proficient collaborators simply by virtue of "growing up" within such structures. Although this may be the case, unfortunately, "hope" is not a strategy for creating collaborative advantage. Future leaders' mere exposure to efficient and effective operating models, matrixed structures, and so on, are not guarantees for success. Collaborative advantage is hard for many leaders to create because it goes against their very DNA and mindset about competition. Power-hungry, self-confident leaders and entrepreneurs have to be very intentional and self-aware about their own defense mechanisms, personal accountability, and unconscious biases, as well as their desire for power, influence, and control—all of which work against effective collaboration. For organizations that must create collaborative advantage to survive, mastering the new demands of leadership collaboration is essential.

Building collaboration competence is necessary for both new leaders and for highly tenured leaders. If you are a seasoned leader who has built a successful career in a command-and-control environment, then you are a collaboration novice. Admit it. You probably haven't seen best-in-class models of collaboration in your career. You've probably experienced plenty of turf wars, hard-fought negotiations, and hostile takeovers, which you might mistake as providing you with an education on collaboration, but make no mistake: successful collaboration is very different than winning a turf war. Admitting that you lack deep collaboration experience and the personal lessons learned from such ways of working is the first step toward getting better at collaborating. If you don't start setting a good example of what collaboration should look like in the future, who in your organization will? Nobody will. That's why this section of the book is so important for seasoned leaders to take to heart.

The ideas and tools in this chapter will help even the most seasoned leaders "unlearn" the hypercompetitive mindset that no longer serves future-focused organizations. Your success with building a collaborative organization depends on your ability to recognize new collaborative ways of work and to model (and teach others to model) them. Likewise, if you are a new leader or an up-and-coming leader, this section will provide you with a framework for creating collaborative advantage and building more collaborative teams.

Collaborative Advantage Requires Upskilling Team Leaders and HR

As stated previously, if your senior leaders fail to model good collaboration, then collaboration won't take root in your organization. But make no mistake: the buck doesn't stop with senior leaders. Good communication and collaboration are capabilities and skills that must be built throughout all levels of smart, connected organizations. Everyone in your organization is going to have to get better at building, maintaining, and managing relationships and power across their collaborative partnerships.

Regardless of your organization's structure, team leaders will play critical roles in helping senior leaders capitalize on their organizational effectiveness/design/ restructuring investment. Team leaders need to develop new mindsets and skills around collaboration to execute strategies and implement priorities set by senior leaders. If collaboration and new organizational design, such as the networked organization, are the new normal, then team leaders are the new "middle management," and they will play critical roles in fostering collaboration.

The "management" practices of previous industrial eras (that is, control, monitoring, disciplining, administrating, and so on) are less and less valuable each day as automation and smart, connected technologies and processes simplify the employee experience. Traditional "personnel management" is becoming obsolete. In fact, I would go so far as to argue that the majority of personnel management practices, and even entire human resources models, of previous eras will soon be obsolete. HR technologies and "employee self-service" capabilities are transforming the roles of HR leaders and HR business partners in more productive and strategic directions. As their transactional work is automated and knowledge work becomes more complex, smart, connected HR leaders will become invaluable resources as consultants and coaches for senior leaders and team leaders on topics related to strategic workforce planning, people analytics, organizational design and effectiveness, leading high-velocity change, and improving team efficiency and effectiveness.

In the last five years alone, I've helped a number of clients "upskill" team leaders and HR leaders, equipping them with a "consulting playbook" and skills around team and organizational development. One role of strategic HR advisers will be to equip team leaders with analytics and coaching for responding to uncertainty and high-velocity change. These strategic advisory and process consulting skills enable functional leaders to "own" leadership and team development more directly within their business unit or department.

My practical experience of working with organizations to build collaboration skills and strategic consulting "horsepower" among HR practitioners is consistent with Richard and Daniel Susskind's predictions in their book, *The*

Future of the Professions: How Technology Will Transform the Work of Human Experts.[3] These researchers argue that we are on the brink of a fundamental and irreversible shift in how the expertise of professions like accountants and management consultants is deployed and used. From firsthand experience, I believe the increased demand to teach the methodologies and analytics of team collaboration, organization design, and so on to corporate function leaders (e.g., those in HR, finance, operations, project management, etc.) is evidence of an evolution of these professions. This evolution is occurring in response to business demand for greater collaboration and more data-driven techniques for transforming human and organizational performance.

Team leaders must learn and master many tactical skills to support an enterprise collaboration strategy. For example, excellent team leadership requires high-impact feedback. This isn't something that can simply be learned over a YouTube-style training video. Great feedback is something that has to be practiced in the moment, during a real business challenge. Because many team leaders struggle to give good feedback, using a trusted adviser as a sounding board and coach is a more efficient and effective use of an HR adviser's time than manually processing forms that a computer could do more efficiently and effectively. The same holds true for senior leaders and team leaders wishing to foster collaboration between matrixed teams, transform culture, work through co-leadership dynamics (e.g., between physician leaders and administrative leaders in health care), and engage in data-driven workforce planning.

To be sure, team leaders will have to master a host of skills to create collaborative advantage. These include things like strategic workforce planning, organization design, change management, people analytics, human–machine collaboration, experience-driven leadership development, succession and scenario planning, and leadership coaching. As organizations build collaborative advantage and undergo digital transformation, systems like Workday, PeopleSoft, and new AI applications will accelerate demand for internal advisory support for team leadership and collaborative advantage. Very few organizations have the type of internal consulting expertise they need to drive efficiency and effectiveness. Building a future-ready organization will require hiring and growing more collaborative leaders.

Why Collaborative Leaders Are So Hard to Find

Some people might be asking themselves, why does creating collaborative advantage require so much teaching, training, and capability development? If you're asking this question, congratulations, you're probably a natural collaborator, and collaboration is just common sense to you. For the rest of us, unfortunately, collaboration is "uncommon" sense. This means that for most people, collaboration skills such as creating shared goals, communicating to set clear expectations, managing conflict and commitments, providing feedback, and so on, must be learned.

Today, it's becoming more common for team leaders to direct many different teams, many in different places and in different stages of team maturity. Many team leaders don't even have formal direct reports anymore. For example, I've spent my entire career in various "team leadership" roles that haven't entailed direct reports. In part, this is because I grew up in consulting organizations. But another reason is because "middle management" for the past 20 years has been slowly winnowed away in organizations. My work as a team leader has always been project-based and has required diverse talents, skills, and collaboration. This means I've had to learn to lead through influence, not position.

Over the past 20 years I've seen similar changes in client organizations. As organizations morph, merge, and grow through acquisition, enterprise work is becoming more project-based and complex, thus requiring diverse talents, new skills, and increased collaboration. As a result, team leadership is becoming more valuable. Of course, one could argue that the team mindset and collaboration skills, as I've described them here, have always been necessary. I would agree with this argument. However, the fourth industrial revolution is exposing the cracks in the collaborative foundation in many organizations and among leaders at all levels. Put differently, the changes to which leaders and organizations must respond require the highest-degree of complex collaboration in the history of humanity. These collaborative demands are pushing the boundaries of leaders' collaborative competence, and many leaders are failing the test.

It astounds me, having worked on hundreds of projects across dozens of companies, how rare collaborative team leadership is. Leaders' mindsets about what it means to be a team leader, to have power, and to effectively collaborate are major challenges for organizations. People's mindsets filter what they believe to be true, fair, and what they consider to be "common sense." Leaders' mindsets also shape how they lead their team(s), what they expect, and what they accept. Some people are naturally inclined to share power and ideas, and enjoy developing others and helping them grow. Some people aren't inclined to do these things. Leaders who want to hoard power, maintain status, and maximize their own rewards and advancement seem to be more common than collaborative leaders. And it's not entirely their fault. When you introduce competition into organizations that have poorly designed incentives, knowledge-sharing processes, and cultures that foster unproductive competition, human nature emerges. Call me a pessimist, if you like. Or call me a realist for speaking to what is plain to see in many organizations in the United States and around the world. Collaboration is the exception rather than the rule.

Meeting the collaborative demands of 4IR leadership requires changing leadership mindsets and organizational cultures. This is the old "nature and nurture" dynamic. In other words, both *human nature* and the *organizational environments* in which people lead contribute to poor collaboration. Data published in the *Harvard Business Review* by Randall Beck and Jim Harter of Gallup help to explain the "human nature" causes for poor team leadership.

According to Gallup research, less than one in five people (18 percent) have "high talent" for managing others at the work group or team level.[4] This means that organizations miss the target 82 percent of the time when putting people in manager or team leadership positions. No wonder organizations struggle with moving to a more complex matrix structure where coordination, shared power, constant learning, and collaboration are critical to success. As this research suggests, most people lack the "collaboration gene" in their God-given leadership DNA.

Another explanation as to why great team leadership is so rare is the organizational environments or cultures in which people work. Most organizations are designed to encourage "upward mobility" and they reward individual contribution. That is, the only way to advance in an organization is to move up or move out. "Moving up" is something that a lot of ambitious people want. Moving up means more challenging work, more responsibility, more decision-making influence, more power, and, yes, more money. The "up or out" culture is the result of a series of misalignments, resulting in bad (i.e., noncollaborative) organizational design.

In an upward-mobility environment, three types of people are promoted to "lead": the best individual contributors, the most tenured, and/or the most ambitious. None of these characteristics have anything to do with who might be the best team leader. In fact, with the exception of the most tenured, these types of people (i.e., great individual contributors and ambitious people) might actually be the worst people to put into team leadership roles.

Let me explain why you should be careful when promoting great individual contributors and/or ambitious people to team leadership positions. First, great individual contributors may not be that interested in helping other people learn and grow. This can be a major barrier to effective team leading and collaboration. The best team leaders want to see others get better in their roles, and they help others develop their skills. Those who are experts, superstar individual contributors at the top of their game, may not have the patience to slow down their own development to help others. It's in the superstars' best interest to keep doing their best work, and to build their reputation as the best in their field. It may also be in your organization's best interest to keep these people in their individual contributor role, for example, as a surgeon, salesperson, engineer, researcher, or designer. You'll get more value from individual contributors who are doing what they love to do and what they are best at, as opposed to making them team leaders or managers.

Some collaboration experts, like Morten Hansen of UC Berkeley's School of Information, would advise executives to fire these superstars, or "lone stars." In fact, in his book *Collaboration*, Hansen argues that leaders cannot make collaborative organizations out of a company full of "lone stars." According to Hansen, organizations should get rid of their "lone stars" and promote T-shaped managers (i.e., those who can work deeply in their own function, the vertical, and cross-functionally, the horizontal part of the T).[5]

However, promoting only T-shaped leaders is problematic for several reasons: (1) not everyone is cut out to be a collaborator; (2) not all *work* requires collaboration (a point that Hansen persuasively argues in his book as well); (3) not all *job roles* require collaboration; and (4) your superstars may be your most highly skilled, technically proficient, and creative geniuses around. Consequently, you could lose a lot of value by firing your lone stars. Just look at Apple: if it hadn't rehired Steve Jobs, perhaps the most quintessential lone star in modern history, after firing him, we might not have the iMac, iPhone, or iPad. The answer lies in designing your organization in ways that allow for individuals to thrive in different roles and to build careers around their greatest talents. I'm not suggesting that organizational structures be built around individuals who might leave, but rather, that people and machines be matched to work that best suits their mindset and work style. At the same time, the workplace environment must incentivize and reward collaborative work where collaborative work is required.

The second reason why I'd caution against promoting great individual contributors to team leader roles is because great individual contributors may also be overly ambitious in their personal career pursuits. Unfortunately, most organizations are poorly designed, such that all roads to increased advancement, rewards, and responsibility go through a stint in management or team leadership. In other words, the only path to leadership in most cases requires that leaders manage people.

Thus, you should be mindful when considering the when, where, and to what team leadership role you're promoting these "ladder climbers." Ambitious people can have the mindset that "it's all about me," instead of "it's all about we." This mindset can kill collaboration. Like highly talented individual contributors, ambitious career drivers may lack the right team leadership mindsets and skill sets to be effective team leads. They may actually cause more damage to a team by doing a two-year "stretch assignment" as a team lead or as a manager of managers.

Senior leaders and HR advisers should really stop and ask themselves, Is leading a team a critical key experience for leading at the next level? If the "management" stop along the way is just a necessary "hoop" to jump through to get to the next pay grade, then consider giving your superstars and career drivers a pass on team leadership and reposition them. Give them special projects that stretch and build their skills. Let them start something new, build something, or turn around something that is failing. But, by all means, don't make them lead a team as a means of retaining them or to tell yourself that you're promoting their engagement or career development! Find something else for your drivers to do to create value and make a difference in your organization that doesn't involve leading a team. Again, mismatched promotions to team leadership are the result of bad organizational and career-path design and ineffective process. This isn't the ambitious employees' faults. It's a career advancement structure and process problem. A better strategy for engaging and retaining ambitious superstar employees is to ask them *what they want out of their career* and help them get there—regardless of whether or not this involves team leadership.

Finally, great individual contributors and ambitious people tend to share similar mindsets about power, influence, and collaboration. These people can tend to be stubborn, self-centered, and even a wee bit ego-driven. Hey, nobody's perfect! These traits probably aren't going to lend to the best administrators or people managers. And that's okay! These very qualities make a great surgeon, gifted artist, amazing architect, bold entrepreneur, or visionary leader; they make them ambitious, talented, and great at what they do. These people are creative and forward thinking; that's probably why you hired them in the first place! So, please, on behalf of stubborn, ego-driven, creatives everywhere, don't make us become managers to get ahead in your organization. Help us target our talents at organizational goals that mean something and align with your purpose.

As demonstrated so far, organizations must get smarter about *who* and *how* they promote people to the ranks of team leader. Smart, connected team leaders—regardless of whether they're born or built—tend to be more patient, amenable, and energized by helping those around them learn and grow. They take pride in coordinating and including others. They may not be the most ambitious "ladder climbers," innovators, creatives, or visionary thinkers, but they are likely better collaborators, connectors, and teachers.

Smart, connected team leaders share power more effectively because they don't have a deep need to be right. My recommendation is to find these people in your organization and put them into team leader roles. Help them grow and advance in their careers too. Provide them with opportunities to teach, coach, and mentor new team leaders. Reward them for great team leadership without expectations of necessarily becoming the next executive leader, P&L owner, or visionary innovator. Smart, connected team leaders who know how to foster collaboration, get things done, and share power are the people who will mobilize the masses within your organization, and who will take your organization into the future. Invest in them. Help them learn new skills in coaching others to be good collaborators, manage conflict, share power, and work with innovators and future-focused drivers. Everyone in a collaborative organization has a unique part to play. Leadership's role is to design and foster collaboration at all levels by finding and positioning the right players in the right roles.

A Smarter Collaboration Framework

So far, I've made the case that learning how to harness the power of collaboration is a leadership imperative for the digital age. Leaders must learn to prioritize work together, manage expectations and commitments, understand their partners' mindsets and priorities for the future, and work across organizational functions quickly to create *collaborative advantage.* Collaborative advantage manifests in the form of sustainable value creation. In the fourth industrial revolution, organizations that can out-collaborate their competition (or simply get out of their own way) will thrive. Others will struggle to survive.

Smart, connected leaders know when to collaborate and when not to. Everyone loves the idea of working together and tearing down "silos," but the reality is, collaboration is an art, not a science. Great collaboration can produce better results, and increase scale and organizational agility. But poor collaboration leads to unhealthy competition, power struggles, and lousy results. Great collaboration, and knowing when to connect teams, requires a smart collaboration strategy.

Let's return for a moment to Professor Hansen's work on collaboration. In his book, Hansen argues that the only way to harness the power of great collaboration is to implement systems of "disciplined collaboration." I couldn't agree with this point more. Smart collaboration is structured, and it is disciplined, with a strategy behind it and structured processes to support it. Smart collaboration helps organizations overcome barriers like information hoarding, poor knowledge management and coordination, power struggles, and turf wars.

This section provides a smart collaboration framework to help you design a more collaborative organization. The smart collaboration framework will also help team leaders foster more efficient and effective collaboration that helps to produce sustained value and collaborative advantage. As I explain the framework, I will provide a set of research-based team leadership practices and tools for improving collaboration and performance, for making collaboration in your organization more disciplined and effective. This smart collaboration framework is also scalable. It can be used at any level of leadership in your organization: from the C-suite to the team leaders in your organization. Before presenting the solution, however, I want to do some collaboration myth busting.

Collaboration Myths

The following are some of the common myths I hear about collaboration. These myths make collaboration even harder than it already is because they perpetuate an unhelpful mindset regarding collaboration. My suggestion is to share these myths with your team, and discuss how, and to what extent, they show up in your organization. When you see these myths perpetuated in your organization, discuss them, challenge them, and bust them.

Tool #12: The Five Collaboration Myths

1. Collaboration always leads to better results.
2. Collaboration is possible without changing how we reward and recognize performance.
3. Old mindsets about competition can coexist in a collaborative organization.
4. Collaborative and noncollaborative organizations use information in the same ways.
5. Collaboration is simply the right thing to do.

These myths come from an ingrained mindset that leaders have about power and collaboration. Because these ideas are so ingrained in modern organizational culture, leaders often take these assumptions for granted. *Collaboration* is a word that has a lot of *positive baggage* associated with it. Leaders assume that "If we just collaborate better, we'll get better results" (Myth #1). However, true collaboration isn't that easy. True collaboration requires discipline. Disciplined collaboration is characterized by collaboration structures, decision-making processes, skills and competencies, rewards, and a different mindset about working together to achieved shared results. Changing all of this stuff is hard work. So, you'd better be sure that collaboration is the right model to achieve your objectives before you jump on the collaboration bandwagon. Some projects and tasks don't require collaboration. That's okay. Know when *not* to collaborate!

Myth #2 states that "collaboration is possible without changing how we reward and recognize performance." True collaboration requires alignment. Leaders must align goals with strategy. Then they have to break down strategies into specific tasks—this isn't as easy as it sounds. As with the adage, the devil is in the details. Once tasks have been established, team leaders have to align expectations around the tasks with the people who are responsible and accountable for them. And, finally, rewards, recognition, and incentives have to support individual and team achievements. That's a lot of alignment! When collaborations lack alignment across all of these areas, from strategy to successful performance, results will suffer.

Myth #3—"Old mindsets about competition can coexist inside of a collaborative organization"—is a tricky one because it's truth or falsity depends on how one defines "old mindsets," "competition," and "collaborative organization." The old mindset about competition that I argue cannot coexist within a collaborative organization is that of the *zero-sum game*. The zero-sum game mindset believes in strict competition, which dictates that for me to win, you must lose.

If I'm operating with a zero-sum game mindset, it is in my individual best interest for me to maximize my rewards and minimize my losses so that, in the end, I win and you lose. A non-zero-sum game mindset seeks alternatives to maximize gains for both parties involved. In a non-zero-sum game, gains can be greater than the sum of their parts. This involves good communication, trust, coordination, and negotiation. Collaborative organizations are those that seek to optimize products, services, and outcomes in ways that produce the greatest amount of good for the greatest number of stakeholders involved. This section will offer skills, techniques, and strategies for fostering this "greater good" (i.e., non-zero-sum game) mindset. Because, as game theory teaches, when one party seeks the "greatest good" but the other party seeks "individual gain" (e.g., using tactics such as betrayal), collaborative results suffer.

Myth #4 states that "Collaborative and noncollaborative organizations use information in the same ways." Collaborative organizations are very good at organizing and using information. Competent collaborators make information open, accessible, and transparent so that optimal decisions can be made to maximize gains. Noncollaborative organizations struggle with sharing information. One surefire way to tell if you're in a collaborative or noncollaborative organization is to ask someone how good the organization is at "knowledge management," "knowledge transfer," or in adopting ideas from outside of their department? If the answer is "not great" or "we really struggle in these areas," there are probably deep alignment problems between what the organization preaches (collaboration) and what it practices ("we don't like it if it wasn't invented here").

Myth #5 is my favorite: "Collaboration is simply the right thing to do." I love this one because it's hard for people to take a stand and say, "I'm anti-collaboration." I, for one, love people who are bold enough to assert this. Make no mistake: collaboration and inclusion are not the same thing. Inclusion is simply the right thing to do. Collaboration is not. Some people, teams, and even organizations don't need to collaborate. Salespeople, for example, are like wild stallions running free in the marketplace. The best sales people are highly competitive, ego-driven, bold, and brash. Don't try to make these wild stallions coordinate their efforts in a bureaucratic fashion. Study your best salespeople. Learn from them. Teach their best practices to new salespeople, and so on. But otherwise, just show them the target and turn them loose to run. And yet, for the sake of argument, I am exaggerating this hypermasculine mindset about competition, recognizing full well that it is becoming less useful in 4IR organizations. Although collaboration isn't always the right strategy to pursue (or the right thing to do), the solution comes down to one of smart organizational design. As such, the most useful thing that 4IR leaders can do is to invest in helping others know when, where, how, and between whom collaborations make the most sense and create the most good for the organization.

Seven Elements of Smart Collaboration

Smart collaboration helps leaders identify and address the collaboration barriers in their organization. The smart collaboration framework provides a rubric against which you can audit the collaboration effectiveness of your organization. You should use the smart collaboration framework to compare and contrast the dynamics and patterns you see within your own organization.

As I've argued, smart collaboration comes down to good organizational design and alignment between the elements that make up an organization's overall design. Richard Burton, professor emeritus of strategy at Duke's Fuqua School of Business, and his colleagues outline a very practical step-by-step approach[6] for executives and MBA students to ensure alignment and/or removal of *misfits* in the organization design process. For Burton, "organizational design in an ongoing executive process that includes both short-term, routine changes, as well as intermittent, larger-scale changes."[7] In other words, the work of designing your organization for collaboration and optimal performance is never finished.

Burton and colleagues introduce what they call the *Diamond Model* to illustrate the elements that make up an organization's design. The elements in the Diamond Model, as Tool #13 illustrates, share a great deal in common with McKinsey & Company's *7-S Model* of organizational design and Jay Galbraith's *Star Model*. These elements of an organization's overall design play a pivotal role in influencing cross-functional collaboration in an organization.

Models of organizational design provide a useful lens to help executives diagnose, redesign, and deploy better collaboration and operations in their organization. The specific elements tell leaders where to look for collaboration barriers. And, perhaps most importantly, these elements tell leaders precisely which levers to pull when making adjustments and alignments to their organization's design.

Table 4.1   Tool #13: Three Models of Organizational Design

Burton et al.'s Diamond Model	McKinsey & Company's 7-S Model	Galbraith's Star Model
• Goals/scope	• Strategy	• Strategy
• Strategy	• Structure	• Structure
• Structure	• Systems	• Processes/lateral capabilities
• Process and people	• Shared values	
• Coordination, control, and incentives	• Skills	• Metrics/rewards
	• Styles	• People/practices
	• Staff	

These three models have some important shared assumptions. First, all three assume that strategy, structure, people, process, and rewards are core elements of organizational design. Second, all three models assume that optimal organizational performance requires that these elements must be in alignment with organizational strategies and objectives. The *why, what, how,* and *who* of an organization's work must all support a common set of objectives. If these elements aren't in alignment, then the organization fails to achieve its purpose. In other words, organizations produce precisely what they are designed to produce. If the organization is designed for excellence, it has the ability to produce excellence. If it's designed for dysfunction, then it will likely produce dysfunction.

And, finally, if you read about these three models in depth, you'll notice some minor nuances in language. For example, "culture" isn't explicitly named in any of the models. However, culture is acknowledged in the form of "practices" in the Star Model, "values and styles" in the 7-S Model, and "people" in the Diamond Model. Similarly, the 7-S Model is the only model to explicitly call out "Systems," which includes technology. Technology is a tacit element of structure and process in the other two models. However, in my seven elements, I have explicitly called out technology as a defining design element of 4IR organizations. Despite these subtle differences in language, the bottom line of organizational design is that these really are the essential elements that drive collaboration, efficiency, and effectiveness.

The most striking difference among the three models is the fact that the Diamond Model explicitly calls out goals and scope. Although I prefer purpose over goals as a higher-level strategic construct, it's important when designing and redesigning your organization for the future to not lose sight of goals, your greater purpose, and the scope of products, services, offerings, and so on, that you exist to offer your customers or constituents. Purpose and goals, ultimately, shape the strategy that you pursue, which are then supported through all the other organizational elements. Therefore, when conducting a collaboration alignment audit, you should always begin by asking, What's the purpose or goal(s) that this organization seeks to achieve?

The Seven Elements of Smart Collaboration Alignment Audit

Having established that the elements of organizational design listed in Tool #13 can significantly impact collaboration and performance, let's look at how you go about auditing these elements in your organization. The *Alignment Audit* can be used to: (1) **diagnose** collaboration enablers and barriers, (2) **analyze** the alignment between the Seven Elements of Smart Collaboration, and (3) **improve** coordination, communication, alignment, collaboration, and—ultimately—organizational effectiveness.

Table 4.2 Tool #14: Seven Elements of Smart Collaboration Alignment Audit

ALIGNMENT AUDIT

Organizational Strategy:	Are these aligned? Yes/No	Unit-level Objective:
Department:	Key Collaboration Partners:	Greatest areas of alignment:
Team Leader:		Greatest areas of misalignment and risk:

Do each of these elements support the others? Are they aligned in supporting organizational strategy?

	Strategy	Structure	People	Process	Technology	Culture	Rewards
Strategy		YES/NO	YES/NO	YES/NO	YES/NO	YES/NO	YES/NO
Structure			YES/NO	YES/NO	YES/NO	YES/NO	YES/NO
People				YES/NO	YES/NO	YES/NO	YES/NO
Process					YES/NO	YES/NO	YES/NO
Technology						YES/NO	YES/NO
Culture							YES/NO
Rewards							

Action Items:

Here's how to use the tool. First identify the goals and objectives of the organization and whichever subunit of the organization you are auditing. For example, if we were to inventory Dimitri's business unit, we would put that unit's name in the "department" box, their unit-level objectives in the upper-right box, and the organization's overarching enterprise strategy/strategies in the "organizational strategy" box. You don't have to analyze every single business or unit-level objective at once. In fact, it may be easier and more illuminating to start with just a single objective or project. Working with specific details of a single project can often surface more immediate collaboration barriers.

Once you've identified these unit-level and organizational strategic objectives, ask yourself the following question: Are these two sets of objectives aligned and supportive of one another? If the answer is yes, then proceed. If the answer is no, stop there and ask yourself why not. Better yet, ask the leaders who are accountable for those objectives whether or not they detect misalignment, and have a conversation about alignment and prioritization. The best way to align objectives is by using the peer-to-peer dialogue from the previous chapter. This tool contains powerful questions for getting to the heart of a misalignment problem with key partners.

Assuming that goals are aligned, the next step is to identify the "team leader" and any "key collaboration partners" in these two boxes provided on the audit form. Key collaboration partners can be individual stakeholders or groups of stakeholders (e.g., other departments, teams, etc.). In Dimitri's case, he might enter Sam as the team leader. Sam is Dimitri's sales manager. And in the key collaboration partners box, Dimitri would enter sales reps, VP of product development, and VP of marketing. Making these collaborative partners explicit is important for accurately assessing the alignment between them in the next step.

The third step in using the alignment audit is simply to circle "yes" or "no" across the various organizational elements. Again, returning to Dimitri's case, at the intersection between "strategy" and "structure," based on what you read at the beginning of the chapter, what do you think he would circle? Dimitri would likely circle a "no" because he doesn't feel that the new matrix structure is supporting his unit-level business objectives. Similarly, Dimitri might also circle "no" where "people" and "culture" cross, based on the fact that historically the different business units have had a culture of competition with one another, and now they're expected to share resources for the good of the enterprise.

At this point, you have an alignment matrix that has some yeses and some no's circled. Okay, now what? The next step is to ask yourself, Among which elements is the alignment strongest and most supportive of unit-level and organizational goals? What makes this alignment so strong? How can we continue to leverage this alignment to increase the speed, acceleration of change, or use it to change our direction? What could other teams learn from our alignment or collaboration in these areas? What might we learn from others to enhance our collaboration in this area? These areas of strong alignment should be written down, communicated, and discussed. For example, if you were Dimitri, you might think that your greatest area of alignment is between strategy and technology based on the successful ERP system implementation that your team achieved with IT and third-party consultants. However, your sales force might not see things the same way. Identify your alignment strengths, discuss them with your key collaboration partners, and correct course.

The next step is to identify any misfits and misalignments. Sit down with your collaborators and discuss, Where are our greatest opportunities to improve alignment? And what risks do these misalignments pose? The answers to these questions are where the real value of this tool is maximized. If you have circled even one "no" on this matrix, it could be a barrier to successfully achieving your objectives and/or accelerating your organization's strategy implementation. If *you* don't fix these areas of misalignment, then who will?

Assuming that you have one or two areas of misalignment, my recommendation is to gather all the stakeholders listed at the top of the form, get them in the same room, pass out copies of this book to everyone, have them turn to Tool #14: Seven Elements of Smart Collaboration Alignment Audit, complete the audit, and discuss their results. Better yet, invite a few stakeholders you didn't write in at the top of your audit form to this meeting. Who has the decision-making authority to align these elements? Who is impacted by the misfits identified by the audit? Make sure these people are in the room. As you're discussing areas of misalignment, don't forget to focus on what *is* working (i.e., the areas of greatest alignment), but don't shy away from the misfits. Address them head-on with *leadership presence* (of thinking, feeling, and acting), and a spirit of problem solving.

To fix the misfits, you have to commit to purposeful actions. Organizational misalignments don't get fixed in a boardroom or at a retreat center. For example, if people in a department have consistently low performance because they aren't using a new system or technology, the misfit problem must be solved where the work is done. Purpose actions could include having managers set clearer expectations, providing training and development to systems users, providing rewards and incentives, or clearly communicating consequences for low performance. Leaders must commit to actions that are specific and time-bound so that everyone is accountable for following up on these issues within a mutually agreed-upon time frame. And, most importantly, leaders must ensure that all actions are aligned with their team's shared purpose and objectives. Most importantly, don't take misfits personally; just fix the misalignment and get on with it.

Maybe you completed the alignment audit, and you don't have any misfits or misalignments. Your commitment to action should be to pull the collaborators listed on the audit together, like I suggested previously, have them complete this audit form, and have some peer-to-peer dialogue with them. Here's how to do it. Send your most important collaborators a copy of this book and let them know that alignment and collaboration are important to you. Let your collaborator(s) know that you'd like to better understand their thoughts on how well aligned they think strategy, structure, work process, and so on, are relative to the areas in which your teams collaborate. Ask them to turn to Tool #14 and complete the alignment audit. Let them know that you are interested in the quality of the collaboration between your teams and how these elements might be impacting that collaboration. Don't tell them that you didn't find any areas of misalignment; just ask them for their candid and honest feedback. Finally, ask your collaborator(s) these two questions:

1.   What areas of alignment make our collaboration most effective?
2.   What areas of misalignment did you identify and how might we address them together?

Inviting other perspectives on these areas of alignment usually surfaces good information, conversation, and can lead to very specific commitments to action.

Deeper Diagnosis of Dimitri's Dilemma

If we turn back to Dimitri's challenge within the new collaborative organizational structure, one way to look at his dilemma is through the lens of the power and relationship dynamics that impact collaboration. These dynamics would fall between the "structure" and "people" elements, or somewhere in the "people" and "culture" space of the Smart Collaboration Alignment Audit.

From the perspective of power and relationships, Dimitri's dilemma is that he feels disempowered to make decisions that impact his business. He was the best salesperson in his business. That's why he was promoted to sales manager and then to general manager faster than anyone before in his company. He knows what his customers and his salespeople need to grow this business! Therefore, he feels that he is in the best position to make decisions that impact his customers and his business.

However, with the new structure, Dimitri now feels like he has lost part of his kingdom. He feels like the organization is preaching collaboration, but in reality, everyone is competing for their piece of the pie. The other leaders with whom Dimitri works (e.g., in marketing, R&D, and finance) have competing "enterprise" priorities, making the work that they do for his unit seem less important. And, finally, Dimitri himself is getting pulled into many different committees, which means more meetings that focus on enterprise-wide work. He feels like he's been pulled away from the job that he loves and that he has done well for his entire career. Lately, Dimitri has entertained the idea of going to one of his competitors and moving back into a sales role—it's less of a hassle and the commissions can lead to a bigger payout than his current leadership position.

In spite of the fact that he doesn't like it, Dimitri feels like he must support the enterprise strategy that the executive leadership team has set. Somehow, he needs to find a way to deal with this dilemma, engage his peers and his sales staff differently, and balance both division and enterprise priorities. This is going to take a lot of energy, influence, and a keen understanding of power relations at the peer-to-peer, team, and enterprise levels. Mastering the new matrixed structure is simply going to require a new way of working and a new way of collaborating.

The Smart Collaboration From/To Table

My diagnosis of Dimitri's dilemma highlights the fact that he feels disempowerment and a sense of loss. Power can be a real barrier to effective collaboration. Good collaboration requires a different mindset about sharing power within complex organizational structures. The language I hear leaders use provides a window into their thoughts, feelings, and actions. And a lot of them have some pretty counterproductive views about power!

Simply put, our words and deeds shape our beliefs about collaboration and organizational changes like the ones Dimitri is facing. Not to get too abstract, but, if you think about it, humans have no way of thinking outside of language. This idea is called "linguistic relativity" and is often referred to as the Sapir-Whorf hypothesis. Linguistic relativity assumes that language influences and, in some cases, controls how we make sense of the world. If you've ever studied a second language, you know that certain words and phrases don't perfectly translate into others. This shapes how we categorize things and communicate about them. Therefore, if leaders don't have the language to describe how power must be shared, and how relationships must be negotiated differently in the future of work, then how the heck are they ever going to get others to collaborate better? Thinking, feeling, and acting like good collaborators then starts with a new vocabulary around collaboration and power.

Table 4.3   Tool #15: Smart Collaboration From/To Table

FROM Power that is . . .	TO Power that is . . .
Competitive	*Collaborative*
Scarce	*Abundant*
Isolated	*Distributed*
Political	*Practical*
Controlling	*Liberating*

Because this is an introductory book to leading in the digital age, I want to ensure that a new vocabulary about power and collaboration is simple and useful for leaders. To ensure simplicity, I've summarized key vocabulary terms in the *Smart Collaboration From/To Table*.

Although the From/To Table might seem like a simple list of words, this table provides a valuable side-by-side comparison between the power mindset that undermines good collaboration (the From list) and the power mindset that supports effective collaboration (the To list). This tool is a great summary of how leaders need to think and behave to get results through collaboration in the future of work, and will help you decide which type of power relationships will serve your organization's purpose and strategy best. As you design your organization to support better collaboration, the From/To Table will help you take stock and test people's assumptions about their power and collaboration mindset.

Use the From/To Table alongside the Seven Elements of Smart Collaboration Alignment Audit to inform conversations with your collaborators about how to improve alignment and collaboration.

Let's dig into some of the assumptions laid out in the From/To Table.

- **From Competitive to Collaborative.** *Smart, connected leadership demands that we evolve from competing for power to collaborating to unleash followers' potential.* Professor Jane Dutton of the Ross School of Business at the University of Michigan has written a great book titled, *Energize Your Workplace: How to Create and Sustain High-Quality Connections at Work.*[8] In this important book, Dutton uses "energy" to describe the power that is unleashed through trusting, collaborative, high-quality connections between people in the workplace. These kinds of connections between people contain the potential power to fuel growth, creativity, respectful engagement, and trust that are absolutely essential for thriving in a 4IR world.

- **From Scare to Abundant.** *Smart, connected leadership demands that power be viewed as an abundant, renewable source of energy, not a scarce resource to be hoarded.* Outdated models of power describe power as something that is limited by the amount of "leverage" that a leader has (or doesn't have) based on their ability to coerce, reward, affiliate, or command their expertise, informational power, and so on. When power is positioned as a finite resource in organizations, leaders waste a lot of energy trying to get their piece of the pie out of fear, distrust, and greed. However, as philosopher Michel Foucault describes it, power is actually something that is accessible to everyone. According to Foucault, *power is everywhere because power comes from everywhere and everyone* (e.g., frontline employees, customers, new technologies, new ways of working, new markets, new products, new ideas, etc.). When power is viewed as something that is self-perpetuating and virtuous, then there is a limitless creative potential between leaders and followers. That's what collaborative power looks like in action. Creativity and possibility unleashed!

- **From Isolated to Distributed.** *Smart, connected leadership demands that leaders let go of the idea that power can be isolated or confined. Instead leaders must learn to accept that power is distributed among followers.* We needn't look farther than the power of social media among employees and customers to see how mighty organizations can crumble under the weight of social connections. Anyone can command a mass audience with the tiny supercomputer in their pocket. As Richard and Daniel Susskind argue in their book, *The Future of the Professions,* even the once common view that "knowledge is power"—something that only "experts" have access to—is changing in 4IR. Successful 4IR leaders will not only accept that these distributed forms of power exist, but they will also encourage it, feed it, and create trusting relationships with followers around the open distribution of power. We already see this in self-managing organizations like Zappos, Uber, and Upwork.

- **From Political to Practical.** *Smart, connected leadership demands that leaders stop playing politics and start getting practical.* If organizations directed even half of the energy that they spend on politics to collaborative problem solving, the problem of global underemployment would be solved. Putting the concerns of followers, customers, and constituents ahead of one's own political and/or financial self-interests is the essence of the kind of "responsive and responsible" leadership that the World Economic Forum has called for to realize 4IR possibilities and potential.

- **From Controlling to Liberating.** *Smart, connected leadership demands that power be free to flow.* Control over ideas, information, knowledge, and potential is something that leaders have less and less of everyday. Control is an illusion. Collaborative power is about letting go of this illusion that you are in control. No single leader is an island. This delusional mindset is the residue of prior industrial revolutions when "labor" was at the mercy of "management." This isn't the case anymore. Today, there are leaderless organizations and self-organizing emergent networks that are radically disrupting stable industries every year. Leaders who accept the shift from leader as controller to leader as liberator, facilitator, and/or moderator of ideas, decisions, resources, and so on, will gain **collaborative advantage** in the 4IR marketplace. That was a really important sentence. I'd suggest reading it again and underlining it!

The assumptions in the Smart Collaboration From/To Table are observable, testable, coachable, and have tremendous practical value. In Table 4.4, I have included a short assessment that will help you measure how you and other leaders in your organization perceive power in your collaborative partnerships. Use this tool for your own self-awareness and to jump-start conversations about collaborative power and potential within your organization or institution.

Table 4.4   Tool #16: Smart Collaboration Power Assessment

Measure	Item	Scale
Collaborative	Leaders in my organization have mastered the matrix structure.	1 2 3 4 5 Strongly Disagree Strongly Agree
	Leaders in my organization know exactly when to collaborate and when not to.	1 2 3 4 5 Strongly Disagree Strongly Agree
Abundant	My coworkers do NOT fear losing power or "turf" in my organization.	1 2 3 4 5 Strongly Disagree Strongly Agree
Distributed	I am empowered to make decisions that impact my work.	1 2 3 4 5 Strongly Disagree Strongly Agree
	Leaders in my organization influence through relationships more than through title or authority.	1 2 3 4 5 Strongly Disagree Strongly Agree
Practical	Leaders in my organization prioritize productivity over politics.	1 2 3 4 5 Strongly Disagree Strongly Agree

Table 4.4   Tool #16: Smart Collaboration Power Assessment

Measure	Item	Scale				
Liberating	Leaders in my company openly share ideas and resources for the greater good.	1 2 3 4 5 Strongly Disagree Strongly Agree				

35 to 31—	30 to 25—	24 and below—
Collaboration Crush	**Collaboration Competent**	**Collaboration Caustic**
Congratulations, your organization is crushing it when it comes to collaboration!	Your leaders are pretty competent collaborators. But there's always room for improvement.	Let's talk. People are actively working against collaboration in your organization.

After taking the Smart Collaboration Power Assessment, do the following:

1. Tabulate your total points for the seven items.
2. Invite someone else to take the assessment and compare notes.
3. If you're a leader of a large team or function, ask yourself how the average employee in your function would respond to these questions? Better yet, send these questions out to your employees and see what they think about how power and collaboration work in your organization! Then have team leads discuss the results with their followers.

The Smart Collaboration Power Assessment can be used to determine a baseline of your organization's mastery of power and influence in cross-functional collaborations. Using this assessment as a basis for discussion and impact planning is an important first step in creating a more collaborative organizational culture. If your organization is "collaboration caustic," then some areas of misalignment likely need to be addressed right away. It might be a good idea to use this assessment to inform your efforts to align the Seven Elements of Smart Collaboration in your organization.

Developing a Collaborative Mindset

The collaboration advice I've offered so far has been fairly "macro" in scope, that is, focusing on teams and departments. I want to offer some ideas for action in this section that are a bit more "micro" in scope, focusing on individuals' mindsets. As you'll recall, "people" are one of the Seven Elements of

Smart Collaboration. There are specific methods of mindset change that can help leaders move their organization from top-down, power-hungry ways of working to more collaborative ways of working. Think of these principles as important first steps toward helping to change individuals and shared mindsets within your organization.

1. Teach your leaders about collaboration and power.

Teaching future leaders about collaboration myths, barriers, and the Seven Elements of Smart Collaboration is essential for changing mindset and behavior. The tools in this chapter are designed so that leaders could teach their teams about these important elements of collaboration and have a productive conversation about how improved collaboration could improve efficiency and effectiveness. Simply having these conversations will change people's mindsets about collaboration. What's more, "unlearning" some of the myths about collaboration can help people better understand that disciplined collaboration means sometimes saying, "no" to collaboration. Not every task, project, or initiative requires intense collaboration. Therefore, learning how to discern when a project warrants collaboration is an essential skill, and part of the mindset change journey. Teaching followers that it's okay not to collaborate on every project or decision reduces the stigma that collaboration is *always* the best way of working. It lets people know that they shouldn't get their feathers ruffled if another team doesn't bring them into a collaboration.

2. Align rewards and recognition.

Most organizations struggle when it comes to shared risk, rewards, and recognition. Smart, connected organizations are "we" cultures. This is a massive shift and struggle for leaders who have grown up in high individualist or "me" cultures. Work cultures are individualist or collectivist in varying degrees by nature. It's either all about "me" in the former or all about "we" in the latter. Knowing what kind of culture your organization has is a powerful step toward changing it.

Aligning rewards and recognition with the collaborative behaviors and outcomes (e.g., improved innovation, efficiency, etc.) is a powerful lever for shifting mindset within your organization. A misfit between rewards and desired behavior will only produce mediocrity and more of the same. Failing to reward and failing to recognize desired behaviors are two of the most common mistakes that organization leaders make when building collaboration. Leaders often resist this feedback from employees, insisting that their formal rewards and recognition programs reinforce the importance of collaboration—*so why don't more people feel recognized and collaborate more frequently?* Most of the time, formal modes of recognition (e.g., award programs, prizes, etc.) aren't what motivate behavioral change. In focus groups, employees often say things like, "Keep your Starbucks gift cards, trinkets, and desktop clutter." What they want instead is for their leaders to notice them, listen to their ideas, and say something meaningful to the people or teams who are truly modeling how the organization needs to collaborate to be successful.

Leaders should make every effort to be authentic in thanking their most collaborative employees. Write a handwritten note versus another email. Call their spouse or mother to tell them how valuable they are to your company. Or better yet, ask them what kinds of rewards and recognition mean the most to them, and use these forms of recognition to thank employees for their difference-making behavior. But for goodness sake, reward and reinforce the behaviors that you want to encourage more of! And, on the flip side, enforce consequences for behaviors that work against collaboration.

3. And finally, remove collaboration killers.

Behaviors that undermine collaboration, such as information hoarding, pursuing individual goals over enterprise goals, defaulting to hierarchical "power moves," and so on, are collaboration killers. If your new structure and operating model require a collaborative mindset among people, then leaders must understand and adopt the right behaviors to model collaborative ways of working. Pinpoint specific behaviors on your team that are killing collaboration and eliminate them. Ask yourself, Who (specifically) is doing what, with what negative impact on collaboration? Once you've identified a recent and specific example of behavior that is killing collaboration, take the following steps:

1. Sit down with and teach the team members the Seven Elements of Smart Collaboration.
2. Describe in detail what you've observed. What happened? What impact did you perceive this behavior having on collaboration and results?
3. Ask team members how they viewed the event? Notice what kind of mindset they have about collaboration? What's their view of power and influence (i.e., more "from" or "to" in nature)?
4. Finally, use future-focused coaching to help individuals change their mindset and behavior. Help the individual set a desired result for improved collaboration. Encourage a specific plan of action that they are willing to own. Establish agreement on the consequences for failing to implement the plan and set deadlines.

Leaders have to *show*, rather than tell, their followers what good collaboration looks like and what they're willing to tolerate. Remember the captain's story about the power of leadership presence? Because leaders are "always on" in terms of setting examples, doesn't it make sense for those leaders to have an intentional plan for developing a collaborative culture and removing cultural barriers to collaboration?

By teaching people the fundamentals of collaboration, aligning rewards, and removing collaboration killers, leaders can create an environment where collaborative mindsets and behaviors take root and grow. This growth won't occur overnight. Although mindset change can happen relatively quickly, true culture change takes time. Usually, the culture transformation of a mid- to large-sized organization takes two to three years. However, these three principles for mindset change are powerful starting points for getting your organization future-ready through collaboration and a renewed mindset about power.

Beyond an Empowerment Mindset

Leaders who do not fully understand power and empowerment from a collaborative mindset might say things like, "I want to *empower* my team to perform at its best." Empowerment and synergy have become clichés because of popular management wisdom. However, *empowerment* and *synergy* are important concepts for understanding how leaders in an organization view power and collaboration. However, these concepts are rarely understood among leaders.

Leaders who demonstrate a more mature, synergistic collaborative mindset about power think beyond "empowering" others. Mature collaborative leaders know that empowerment, by definition, implies that leaders "have" the power to "give" to followers. What's balanced about that? Nothing, the empowerment relationship dynamic is an unequal relationship where one person has power and decides to "give" it to another person. Empowerment is hierarchy with a sweet sugar coating on it. The empowerment mindset goes something like this: "I'm still the boss, but I want to empower you. So now do what I say." If a leader operates as if they have power to give, then they're still modeling a fairly naïve power mindset.

The most mature, synergistic mindset about collaborative power sees power as a *by-product* of a collaborative process. In other words, power (for a team or organization) emerges out of the team collaboration and communication processes. Therein lies the *synergistic power* of true collaboration. Within a synergistic mindset, leaders don't have power to give because power is created out of the interactions between people. Viewing power as an interaction, a process, or an energizing conversation among many people changes the entire relationship between leaders and followers. However, to adopt a power-as-process mindset, one has to let go of one's need to control. You also have to let go of your need to "be the boss." You have to coach yourself (or get a coach) to help you become humble, appreciative of others' talents, and secure in not being the smartest person in the room.

People, especially millennials, want to follow these kinds of leaders. Leaders with this mature collaborative mindset view their roles as one of connector, facilitator, trust builder, inspirer, thought partner, talent advocator, barrier breaker, and team builder. Which of these words best describes you?

Transforming Mindset Thought Experiment

I want you to practice applying the steps in the last section to help Dimitri transform his mindset about the new structure and struggles he's encountering. What would you teach Dimitri about his mindset regarding power (tip: use the From-To Table)? What advice would you give him for rewarding and recognizing collaboration in his business unit (e.g., between sales, marketing, R&D, and other enterprise leaders in operations, finance, and HR)? How would you guide Dimitri in his dilemma using the collaboration killers coaching model I outlined? What behaviors would you encourage Dimitri to stop, start, and continue doing?

This might be a good thought experiment to try with your leadership team. Give your team a copy of the Dimitri case study, and ask them the following questions:

1. What kind of power is Dimitri accustomed to (competitive or collaborative)?
2. How has the new matrixed organization changed the relationships of power? What kind of mindset about power and collaboration currently exists? What needs to change for more synergy and collaboration to occur?
3. What new mindset or skills do leaders like Dimitri need to develop to effectively collaborate in this new world?
4. How is Dimitri's dilemma similar to or different from how our organization operates?
5. What are a few small steps our leadership team can take to improve collaboration?

If you try this thought experiment with your team, here are a few things to emphasize to get the most out of the conversation with your team:

- Dimitri is feeling a loss of legitimate authority to make decisions.
- Dimitri wants to grow his influence by being seen as an "enterprise leader."
- Dimitri's ability to influence through his position and rewards (commission) has been reduced by resource allocation in the new structure. How, then, should he motivate and engage his sales force?

If Dimitri wants to maximize collaboration, he has to consciously change his mindset about power and influence. Instead of approaching power as a fixed commodity, Dimitri needs to start viewing power as a process and relational source of energy, creativity, team engagement, and productivity. He might need to insist that a leader from his sales force be invited to enterprise strategic planning meetings. He might need to prioritize advanced analytics and forecasting within his business unit to procure enterprise resources to fuel growth. This isn't necessarily competing for resources; it's being a good business leader and partner to his enterprise peers. Being future-focused on data projects, innovation opportunities, and market forecasts and alternative scenarios will allow the enterprise to pursue smarter, more data-driven growth paths, and make smarter business decisions.

Advice for Dimitri using the collaborative approach to power and influence:
Again, if you're using this case study as a point of discussion with your team, keep these thought starters and recommendations in your back pocket:

- Try changing your mindset about power—it's not a scarce resource. Power is a process, not a product. Consider this: What would happen if teams worked with a start-up mentality? What if the possibilities for growth were limited only by the speed of innovation, communication, and collaboration of the team?

- Try connecting salespeople with people in R&D, marketing, and other functions whose work might benefit from their perspective—and vice versa. What would happen if we stopped resenting those with other functions for slowing down progress, and started with the question, How can I be helpful to you? This mindset is very different than how most people approach cross-functional collaboration.

- Try developing team-based, enterprise incentives. What would happen if organizations started rewarding cross-functional teams for *actually* collaborating? What if companies put their money (rewards) where their mouths are (e.g., around collaboration and innovation)?

- Try talking with the sales force about the value proposition of the collaborative, matrixed structure. Organizations don't move to collaborative structures unless there is business value for the entire enterprise. Collaboration is too hard. If there isn't business value or value to customers in collaborating, then why do it? What if companies were transparent about the business case for cross-functionality? What if they used this business case to inspire a sense of ownership, commitment, and enthusiasm for learning how to go farther and faster together?

What additional explanations and advice can you come up with? How might the people with whom you work or collaborate make sense of this case? Maybe you should ask them!

Cultivating Collaborative Partnerships

Moving from the most "micro" level of mindset change, I want to provide you with some resources for creating smarter connections and partnerships. The "partnerships" to which I'm referring are the one-to-one relationships that make up your organizational network. Leaders who are great at creating powerful partnerships understand that power is something that resides within the connections in their network, not with individuals. Power exists where two people or two forces work together. Power always involves a relationship between two forces, such as past–future; stability–change; predictability–uncertainty; depth–breadth; strategy–structure; people–process, and so on. Researchers call these "relationship dialectics," and smart, connected collaborators can spot these dialectical tensions and know how to communicate through them to create shared futures and results.

Think back to Dimitri's case. He is juggling many different relationships within the new organizational structure. There are different relationships of power and different dialectics at play between each group and the partnerships that exist within each group. Figure 4.1 illustrates Dimitri's collaborative relationship map based on the case at the beginning of the chapter.

Dimitri's Collaborative Relationship Map

The quality of each of the relationships that Dimitri has with his peers represented on this map is where his true collaborative power exists. As the case study states, with the change in structure, Dimitri's power shifted from his title and position to his network of influence.

Within this constellation of relationships, one of the tensions or dialectics that Dimitri must master is with his peers. In the old structure, Dimitri only had a handful of peers—these were the people who led the other three business functions. He had very deep, trusting relationships with these people with whom he had worked for many years. Now, in the new structure, Dimitri has *many* more peers who have influence on his business. The tension created by moving from a few deep relationships to a broader constituency of many peers illustrates the relationship dialectic of *depth–breadth*. This tension changes the demands of Dimitri's role in that he has to make more "friends." Specifically, Dimitri has to learn how to build trust, establish that he too is trustworthy, discover what kind of mindset and values his peers have, and explore how to best partner with his new peers. Managing depth–breadth, at a minimum, is something that Dimitri has to be aware of. In the best case, Dimitri would manage the depth–breadth tension with his peers in such a way as to maximize the creative power that exists within each of his new peer-to-peer relationships.

A relationship-centered view of collaborative power reveals that, just like Dimitri, 4IR leaders do not have "power over" other people. Leaders now

Figure 4.1

have to manage "power with" their followers, peers, and constituents. When we have higher-quality relationships, we have a greater likelihood of generating more power and creative energy with those people. The first step in cultivating smarter and more connected partnerships is to inventory the people with whom we have relationships, and to assess the strength of those connections.

Tool #17 is called the Collaborative Power Relationship Map. This planning tool will help you map and build greater influence with the people in your network. Fill out the following relationship map and add as many extra bubbles as you need. Include groups of people with whom you have formal and informal connections at work. Include people with whom you have high-quality and low-quality relationships.

Who are the individuals with whom you have the most power and potential? With whom do you have the highest-quality relationships? How will you leverage your high-quality relationships and repair your low-quality relationships? What untapped potential exists in this network of power

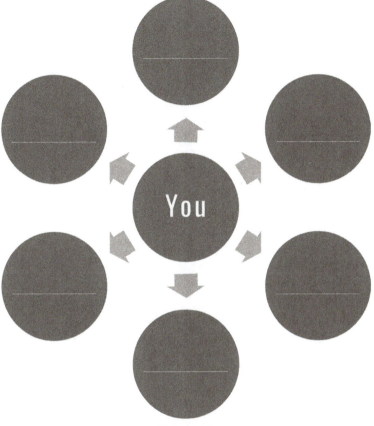

Figure 4.2

relationships? What potential are you leaving on the table? What new relationships do you need to create?

Tool #17: Collaborative Power Relationship Map and Action Planning Guide

Once you've completed the relationship map, create a plan. The following questions will help you create a plan to maximize the collaborative power that exists within your network.

1. What is the current status of each relationship? Use a good, better, best rating.
2. What can I invest in: building more trust, sharing more information, repairing a strained relationship, or making a good relationship even better?
3. What relationships are missing from this network? What relationships do I need to build?

4.   Whose relationship map might I be on that I'm not aware of?
5.   What's my accountability plan? What action will I take? What's my time line for completing these actions? Who will help hold me accountable for results?

Building higher-quality relationships is the key to mastering collaboration in a 4IR organization. Leaders can no longer be successful with using coercion (sticks) or reward tactics (carrots) alone to influence people. The best leaders in the smartest and most connected organizations truly recognize that power is not something that they have "over" followers; it something leaders create through collaborative communication and interactions with followers. If you take nothing else away from this chapter, my hope for you is that you take this lesson forward.

Chapter Summary

1.   Fourth industrial revolution volatility and disruption requires greater collaboration and shared power among leaders and organizations. This chapter has focused on building collaborative advantage between functions within an organization.
2.   In this chapter, we've established why collaborative leaders are hard to find. We've also debunked some collaboration myths and addressed some of the barriers to effective collaboration.
3.   This chapter also took a deep dive into the Seven Elements of Smart Collaboration and provided a framework and tools for auditing these elements and improving the alignment among these seven elements in your organization.
4.   Finally, this chapter introduced six collaboration-enhancing tools: the Five Collaboration Myths; Three Models of Organizational Design; the Seven Elements of Smart Collaboration Alignment Audit; the Smart Collaboration From/To Table; the Smart Collaboration Power Assessment; and the Collaborative Power Relationship Map and Action Planning Guide.

Recommended Actions

1.   Take the Smart Collaboration Power Assessment, and have others take it too. Then discuss your results.
2.   Complete the Collaborative Power Relationship Map and Action Planning Guide and create a plan. Share your map with everyone in your network with whom your collaborative power exists. Share the Dimitri case study and relationship map if trust on your team is not yet high enough to do the exercise among yourself and your team.
3.   Finally, use the Seven Elements of Smart Collaboration Alignment Audit to start building a more collaborative, innovative, and effective organization.

CHAPTER FIVE

Development

Nihil is a software engineer working on a very important artificial intelligence pilot project. Six months ago, after completing his degree, he moved to Des Moines, Iowa, to start his first job. Nihil is working with a global manufacturing company, helping them make smarter products that help the agriculture industry. Although Des Moines isn't exactly "the Bay Area" experience that many of his college buddies are living, he's enjoying his new job and his new home. His starting salary of $94,000 goes a lot farther in Iowa, and he enjoys the active road biking community in his spare time.

Nihil describes himself as an "always-on" learner. Anyone who has worked with him would absolutely agree. Since he was a little boy, Nihil has spent several hours per day on the Internet, reading articles about new technologies and coding. In school, he was one of the most creative AI programmers in his class. But Nihil doesn't solely focus on learning new technology skills. He spends a lot of time watching YouTube videos on Japanese watercolor painting and a 17-string instrument called the koto. On the weekends, he takes virtual koto lessons from a teacher he met online who, ironically, lives in the Bay Area.

When Nihil's team leader, Sarah, found out about his hidden artistic talents, she asked him how he finds the time to do all this. Nihil said, "If I don't make the time to study creative disciplines like painting and music, then my coding suffers. In order to understand how machines learn and develop, you have to constantly learn and grow too."

Sarah is neither a great programmer, nor is she an artist or musician. She is, however, a great team leader. Although Sarah doesn't quite understand the connection that Nihil sees between machine learning and watercolor painting, she wants to support his ongoing growth and development. Nihil is a superstar and Sarah understands that millennials like him want to work for companies that provide ongoing learning and growth opportunities in work and life. In just six months, Nihil has proven to be an amazing member of her team, and so Sarah wants to keep him (and others on her team) focused, motivated, and engaged.

In doing some research of her own, Sarah learned about how "smart creatives" and "always-on" learners like Nihil are motivated and how they think about problems. Some research Sarah found demonstrates that always-on learners have a different mindset about learning. Sarah also learned that smart creatives are not always the easiest people to manage! Their interests change, they have wacky ideas, and they come up with passion projects they want to pursue. Through her research, Sarah found that Google has learned a great deal about how to lead and empower smart creatives for development and professional growth.

One idea that piqued Sarah's interest came from Google founders Larry Page and Sergey Brin. In their 2014 initial public offering (IPO) letter, Larry and Sergey publicly communicated their philosophy of "20 percent time." This is a management philosophy that states that employees should be allowed to spend 20 percent of their time on creative projects they think will benefit Google. By empowering employees to be more creative and innovative, 20 percent projects have produced innovations like Gmail and AdSense for Google.

To prepare for her development planning conversation with Nihil, Sarah also watched a few TED Talks on how to manage the workforce of the future, which provided her with tips for improving her management and team development skills. Finally, Sarah discovered a few free industry research reports (*like the ones that will be reviewed in this chapter*) and podcasts that she downloaded to her personal iPad and reviewed while she was traveling on business.

Welcome to the new culture of learning and development that is driving the future of work! Continuous and highly connected learning is becoming an essential strategy for individual, team, and organizational development. Complex knowledge work that maximizes human creativity, critical thinking, and emotional intelligence is only possible if organizations invest in the lifelong growth and development of their people. Compliance, credentialing, and "nice to have" training just simply won't get the job done in the future of work. People (and organizations) who struggle, or who are not motivated, to learn and grow will fall behind in the fourth industrial revolution.

As this case illustrates, the "learn or become obsolete" mantra is as relevant for technical employees, like Nihil, as it is for team leaders like Sarah. Staying on top of the mega trends that are disrupting products, processes, and organizations requires continuous learning. And those organizations that make it easier for people to continuously learn, hone their skills, and develop new skills will prosper.

Throughout this book, I've provided readers with tools for developing themselves, their teams, and their organizations around topics related to *presence, agility,* and *collaboration.* This section is about *development.* I contend that future-ready organizations have leaders at every level who know how to develop the workforce of tomorrow. Unfortunately, many leaders don't seem to understand that development is a fundamental requirement of their job. In their defense, that's not entirely their fault. Executives who lead many organizations often fail to prioritize development with any degree of strategic design, sophistication, or accountability. Let me be clear: in the majority of organizations I've observed, development is inconsistent, lacks strategic focus, and is woefully inefficient and ineffective. Even among the small minority of leaders who understand that developing employees and staff is a requirement of their job, many of them are not proficient. Building a future-ready organization requires that leaders start taking development more seriously and start investing in development more strategically.

The role of learning and development in organizations is changing. Industries, organizations, and entire professions are becoming highly specialized and even fragmented. Every knowledge-driven profession is experiencing this change. Work is becoming more interdisciplinary and complex, which means that people need to learn how to do *more* and *different* kinds of work each year. Amid the ongoing need to develop highly skilled knowledge workers, it's absurd to think that your organization's HR or learning and development function has the capability to be the sole stewards of this learning and development.

HR and learning and development specialists can support leaders in designing learning experiences based on strategic organizational goals, but these professionals are by no means experts on every subject matter. Future-ready organizations embrace the idea that development is everyone's job—including the individual responsibility of your staff and employees. To be successful, future-ready leaders and organizations must have a development strategy, goals, metrics, and impactful processes for cultivating individual, team, and organizational development. The success of their organization lies in the balance of equipping leaders with the knowledge and tools for developing their employees and teams. Therefore, leaders must start improving their skills at developing people now. As one client shared with me, "Iron sharpens iron. The only way to develop a leader is to have another leader show them the way."

Simply put, in future-ready organizations, leaders at all levels have the knowledge, tools, skills, motivation, and incentives to develop their workforce. Future-ready organizations are designed to support learning and development through both informal and formal means. Clients always ask me, "Why do some organizations outperform, learn, innovate, engage, change, etc. other organizations?" My answer is simple: because the best organizations are designed for excellence. Excellence is a choice, and so is mediocrity—either is up to you. You get out of your organization, and your people, precisely what you put in. Some organizations choose excellence by taking the necessary steps to design for excellence and develop capabilities (both technologically and in their people), and other organizations choose mediocrity. In the future of work, choosing mediocrity is the same as choosing obsolescence. Therefore, putting together and implementing a future-focused development strategy is a must for future-ready organizations choosing excellence over mediocrity.

One reason why smart, connected organizations are, well, *smart* is because learning and development are integrated into "the work" itself. In other words, people get paid to not only *do the work*, but also to *reflect on* the work, *learn* from the work, and *redesign* the work in ways that add value to their organization. Learning and development are woven into the very fabric of how work is done, how problems are solved, and how business decisions are made. Designing a smart, connected culture of development and growth is radically different than the "training and development" approaches that most organizations have used in the first three industrial eras. Choosing excellence in learning and development doesn't mean choosing a better technology system or curating better training content, it means *choosing a better strategy* for developing people, teams, and the organization. Sure, your development strategy might require better courses or training materials, but these elements are simply the means to achieving development ends, not an end in and of themselves. Don't confuse the means and the strategic ends of your development strategy.

Old approaches to "managing" learning and development with an out-of-date learning management system are broken and obsolete. Most organizations are better off scrapping 80 percent of their competencies, training, and learning management system investments and just starting over with a strategy that works. Training is a path to average. Excellence requires deep development.

Smart, connected organizations of the future will be led by executives who understand the basics of learning and development on a deeper level. These leaders will know how to teach, coach, and guide people and teams through complex on-the-job experiences and challenges. By teaching and coaching, development-savvy leaders will help their people make sense of on-the-job experiences, sharpen their skills, and align their passions, talents, and job tasks with their organization's strategy and purpose. This is no small task. It will require that leaders understand the nature of what people actually do. It will require that leaders understand deeply where the organization is going and what it's trying to achieve. And finally, it will require that leaders are able to "connect the dots" between their teams' daily tasks and the organization's direction. Thank goodness, future-focused organizations are already discovering best practices in building cultures and communities of deep development.

In these "best practice" organizations, continuous growth and development of people are outcomes of a highly integrated learning culture. Schools, organizations, and companies that are fostering deep development put a premium on how people learn, grow, and develop. And because development is valued more in these learning organizations, they "do learning" differently. Learning and development are organized differently. For example, instead of learning and development being one person's job, that is, the head of HR or "corporate learning," learning is everyone's job. Best practice learning organizations use communities of experience and communities of practice to foster deep development at all levels.

Communities of experience are groups of people who share similar interests, have common goals, hang out in similar locations (virtual or physical), and participate in shared activities. Employees in the same organization, field, or project team are all examples of a community of experience. Similarly, customers within a particular segment of your business are another example of a community of experience. The community is simply defined by a shared set of experiences.

Smart, connected organizations intentionally foster communities of experience around shared learning experiences. The interactions and communication within the community allow deeper learning to take place and extend beyond initial exposure to new experiences or ideas. The communication among community members supports ongoing growth and development that can be easily shared, transferred, and integrated into the organization's products and services via the community.

For example, one of my clients, a consulting firm, has a health care community of experience. This community is made up of all kinds of people who have the shared experience of working with health care customers. Within this community, consultants who work with health care clients—regardless of their role, rank, or level of expertise—can share experiences and lessons learned, pose questions, and submit ideas for improving the company's offerings to health care clients.

This community of experience is held together through monthly learning and development meetings (e.g., around safety, value-based care, coaching, developing physician leaders, etc.), and through continuous technology-enabled conversation. Prior to and following formal meetings, the health care community of experience is always connected around industry news, research, and leading practices emerging from their experience in the field. This is a great example of how smart, connected, and committed employees have self-organized around a topic that they are passionate about, and that aligns with their consulting firm's strategy. It also highlights the importance of individual responsibility and one's own responsibility to invest in deep development outside of their "normal" job duties.

Best practice learning organizations also foster communities of practice. **Communities of practice** are made up of practitioners of a particular profession, craft, or hobby. These are experts within their field, niche, or industry. The community is made up solely of enthusiasts who seek deep knowledge and shared learning experiences in pursuit of common goals. On the surface, communities of practice sound a lot like communities of experience. But there are some important differences.

For example, within the health care community of experience, described previously, a small subset of health care practitioners (around seven) are leaders within the broader community of experience. These individuals are truly health care enthusiasts and experts. They have deeper knowledge than other members of the community. They are influencers within the community without formal authority. They work outside of the broader community of experience to seek new knowledge and set the learning agenda for the broader community. They are thought leaders and opinion leaders within the community of experience. These deep development gurus have separate meetings on defining the future direction for the firm's health care practice. They are not leaders by title; they lead through their purpose, passion, and expertise.

These two examples of how smart, connected organizations foster deep development help to set the stage for what this chapter is all about: facilitating deep development into your future-proofing strategy. In this section, I provide important data on the changing nature of workplace learning. I also outline valuable tools that will help you develop your organization for the fourth industrial revolution challenges and opportunities that lie ahead. Let's start by examining why development must be approached differently in future-ready organizations.

A Solution for Our Broken Approaches to Development

The way that most organizations approach learning and development is broken. Despite the fact that organizations have access to more information at the click of a button than ever before in human history, we have a human development crisis on our hands. This development crisis is global in scope, and is associated with huge economic risks. As I've already argued, 4IR carries with it many pitfalls and economic risks stemming from automation, labor shortages in high-skill fields, social inequalities, and IT security threats.

Given these risks, we should be asking, What role does deep development play in avoiding 4IR pitfalls and maximizing human potential? The short answer is that development is hugely important. To maximize potential, leaders must expect more of themselves and their organization's approach to human development. Success in the future of work will demand that organizations get better at cultivating lifelong learning and development among staff, employees, and their constituency. Success will also require leaders to honor and celebrate development as an essential element of their social and functional "contract" with employees. In other words, organizations must design learning and development into every element of their architecture (see Tool #13: Three Models of Organizational Design).

Historically, organizations have been run solely as places where work is done with limited learning, development, and growth taking place. These organizations were characterized by "shallow" forms of human development. Shallow forms of development are *superficial and transactional in nature, resulting in incremental gains in knowledge or skill* on an as-needed or compliance-driven basis. Regulatory and safety trainings are examples of such shallow development. Continuing education and professional credentialing are slightly deeper forms of shallow development that may or may not result in learning that can be applied to improve performance. Figure 5.1 illustrates examples of development activities that range from shallow and transactional to deep and transformational.

Types of Shallow versus Deep Development

By contrast, deep development is a *transformational process for learners that results in mindset change and enhanced capacity and motivation to change the world around them*. This includes things like breakthrough, on-the-job experiences

Shallow/Transactional Development

Regulatory

Compliance Training

Classroom Learning

On-the-Job Lessons

Passion Projects

Deep/Transformational Development

Figure 5.1

that result in important lessons learned (e.g., like the captain's lessons around presence of thinking, feeling, and acting). Deep development could also include self-paced learning on an applied "passion project"—like Google's 20 percent time. Deep development could also include formal learning journeys (e.g., an off-site retreat for executive development) that involves intense coaching focused on mindset change. Using the left-hand side of Figure 5.1, write in some examples of development programs or activities from your own organization. Are the majority of your learning activities deep or shallow? Are they transactional or transformational?

The role of deep development in organizations is changing. Organizations are not simply places where work is done; they're places where people learn, develop, and fulfill their life's purpose. If organizations expect the best, most creative and innovative thinking from their staff and employees, then they had better embrace the fact that human beings need deep development. The imbalance that has existed between expecting deep results while only offering people shallow development is unsustainable. Organizations and their leaders must improve deep development if they want to optimize the creative potential of humans in the future of work.

Improving deep development means truly understanding and implementing a deep development strategy, not simply making incremental investments in "training and development" and calling that a strategy. As stated previously, such strategies are shallow development, and shallow development only produces mediocrity. Fourth industrial revolution challenges should serve as wake-up calls to leaders that it's time to better use the billions of dollars that their organizations waste on shallow development.

The Five Principles of Deep Development

The Five Principles of Deep Development offer a helpful framework for clarifying why deep development is the best solution to fixing broken approaches to learning and development. I've summarized the Five Principles of Deep Development in Table 5.1 and will briefly discuss them.

Table 5.1   Tool #18: Five Principles of Deep Development

Principle	**Shallow Development**	**Deep Development**
Outcome	Transactional	Transformational
Delivery	Episodic	Continuous
Integration	Disconnected from work	Integrated with work
Design	One size fits all	Fit for function
Strategy	Unintentional	Deliberate

Principle #1: Deep development is transformational.

Deep development changes adult learners involved. To use the phrase of Harvard University's Robert Kegan and Lisa Laskow Lahey,[1] deep development is *vertical* rather than *horizontal*. The outcome of deep development is a mature, self-authoring mindset, not just an expanded skill set. This is not to say that the future of work doesn't require horizontal or transactional skill development—it absolutely does. But current professional development and education models are pretty good at transmitting information and building skill. Anyone can watch a few "how-to" YouTube videos, but truly transforming *how* someone does work (not just *what* they do) takes a different methodology altogether. Learning and producing extraordinary outcomes requires the support of a community of experience and a coach or facilitator to help you make sense of new developmental experiences, and most of all, time to take root.

Principle #2: Deep development is continuous.

Transactional or horizontal development is episodic, whereas deep development is continuous. To illustrate the broken model of development that most organizations have, simply look at the language they use. Phrases like *self-paced* and *just-in-time* learning have become popular vocabulary within organizations. HR people talk about *self-paced* and *just-in-time* learning like it's a badge of honor or some remarkable accomplishment they've pulled off by spending millions on a new "learning management system" to support shallow learning in the nick of time. This is not something to be proud of. It's development failure and the normalization of that failure. Although just-in-time applications of knowledge and skills are consistent with some principles of adult learning, they aren't a sustainable strategy for mindset change or mastery. They're strategies for shallow, transactional learning.

Let me put it in these terms: Do you want a surgeon who completed a "just-in-time hernia module" to perform your surgery? Do you want an accountant who completed a *self-paced* online degree over seven years to do your financial audit? Do you want to pay for a university education taught by professors who've hand no expert mentors or advisers, but simply completed a series of self-paced content modules in an area of interest? Do you want your 16-year-old's car fixed by a mechanic who just learned how to change a tire from a YouTube video? Of course not! Similarly, you don't want leaders who don't understand deep development making important business decisions or managing the smart creatives in your organization.

Developing expertise takes time and is built over a continuous process of horizontal and vertical development experiences. Although self-paced and just-in-time learning might work for shallow development, it doesn't work for deep development. Stop kidding yourself that your "just-in-time" strategy is adequately preparing your organization for the volatility, complexity, and high-velocity change of the fourth industrial revolution. A future-ready organization must have a strategic balance between well-curated, shallow development (self-paced) content offerings, but also a strategy for promoting deep, continuous development across employees' career life cycles. Therefore, organizations must learn to deliver deep developmental experiences in an engaging manner over the course of their employees' careers to achieve the transformational outcomes and leadership capability they need to succeed.

Principle #3: Deep development is integrated with work.

Deep development requires context, immediacy, and situation-specific decision making. Classroom learning is great for elaborating and discussing new knowledge, concepts, or skills. This requires great trainers, educators, and facilitators. However, at some point, classroom knowledge and skills have to be put to the test in the real world. If knowledge and skills are kept in isolation from day-to-day experience, then the best possible outcome one can hope for is incremental growth in knowledge or transactional learning.

Deep learning requires integration with the experiences of everyday leadership. It requires real-time (and in some cases "just-in-time") application, over-the-shoulder coaching, and advice from peers or team leaders. Therefore, because deep development must be integrated with work, leaders at every level need to have basic coaching, advisory skills, and frameworks for making sense of new real-world challenges. These are fundamental, deep development skills to help tomorrow's leaders gain the wisdom and experience to manage complexity in the future. HR professionals don't necessarily have the domain expertise to provide content-specific advice and coaching per se, and even if they did, most organizations don't have enough HR or learning professionals to go around. Consequently, leaders at all levels need to be taught how to think critically, ask good questions, and coach one another regarding key experiences that contribute to deep development on a day-to-day basis. An effective deep development strategy leverages the HR function to build these coaching capabilities at all levels. Moreover, HR functions that support deep development use data-driven models of on-the-job experience to measure developmental growth, maturing, and return on development investment (RODI).

Principle #4: Deep development is fit for function.

Whereas training models of the past are one size fits all, deep learning strategies are fit for function. Broad-based competencies can be effective, but nine times out of ten, they're just cumbersome HR speak. Deep development organizations have leaders who know how to tailor a learning plan for their teams and for specific individuals. These development-minded leaders know what kind of vertical and horizontal development their people (and the organization) need to be successful and to procure the resources and support to help them get it. In addition, development-minded leaders individualize coaching and feedback to their staff in such a way as to maximize the learners' growth and development. Finally, leaders who know how to promote deep development make smart investments in their people based on the current needs of their organization, and the future needs a particular person might be able to fill.

Principle #5: Deep development is intentional.

Deep learning has a strategic intent and is not a hodgepodge of content delivery for the sake of "learning and development." As organizational strategy and needs change, so too must deep learning strategies. Content must be continuously evaluated, assessed, and refined. Organizations need to learn how to make smarter and more sustainable investments in development analytics and evaluation of RODI. If investments in learning "spa days" and top-dollar Harvard executive development weekends aren't producing value or change in leaders, the programs should be abandoned, and new strategies pursued. In the United States alone, organizations spend in excess of $70 billion on training that fails to produce RODI. These kinds of investments could buy a lot of great deep development, but those responsible for training budgets are rarely held accountable for demonstrated business outcomes. In the future of work, organizations must develop strategies, structures, and processes, and place the right people to lead intentional, strategic, deep development programs.

These five principles are a great framework for reflecting on and guiding for improved development in your organization. They can be used to evaluate your organization's ethos and collective mindset about learning development, and to take the first steps toward fixing them. In what follows, I'll share more data to make a case for deep development in your organization. And I'll build on these principles with some specific strategies and tools for overhauling development to build the organization of the future.

Using Deep Development to Create a Future-Ready Organization

Employees like Nihil and his team leader Sarah, mentioned at the beginning of this chapter, are the new norm in knowledge-driven organizations. Therefore, you have to start preparing your organization with a deep development strategy now. But the trouble is, most people don't know where to start in overhauling their development approach. If I've learned anything from helping leaders tackle disruptive change, it's this: all organizational change begins with data, awareness, and a basic understanding of *why* the current state isn't sufficient for meeting demands of the future. Once people understand the why, they can usually figure out the *what* and the *how*. Thank goodness, there are some very useful data to help you explain the why behind deep development to "the powers that be."

According to Bersin by Deloitte's 2017 *Global Human Capital Trends* report, which surveyed 10,400 executives in 140 countries, 83 percent of respondents said *careers and learning* were "important or very important" topics. In fact, learning and careers were the second most important trend listed in the Bersin report. Here's another encouraging sign: the learning and careers trend, according to Bersin, was second only to focusing on building the "organization of the future."

Deloitte's research also shows that 90 percent of CEOs agree that their companies are facing digital disruptions. Of those CEOs, 70 percent feel that their workforce *lacks the skills to adapt*. This is staggeringly low confidence in the workforce of the future. So, if you've been tracking along, Deloitte's data show that:

1. Executives believe that their organizations are facing massive digital disruption, and that building the organization of the future is their top priority.
2. But the majority of executives admit that they don't know how to build the organization of the future, and that their workforces also lack the skills to adapt to this change.
3. So, the executives that Deloitte surveyed are focusing on two major areas for stomping the flames of this burning platform: careers and *development*.

The logic behind focusing on learning and development makes sense, given what we've covered thus far. But this is where the story gets a little wacky. In surveying 800 top business executives, Deloitte also found that two-thirds (67 percent) felt *technology* rather than *human capital* will drive greater value for their organization. In addition, 64 percent reported that people are a cost, not a driver of value. In other words, what senior executives think about the role of human development, technology, and their respective value in the future of work seem to be at odds.

Collectively, these data show that the vast majority of top executives don't understand the workplace of the future or how value will be created through people AND technology. Or, perhaps, executives' waning confidence in human capital is the result of poor past performance? Incidentally, Bersin research finds that the majority of non-HR executives (70 percent) report that their organizations' HR capabilities are inadequate.[2] As such, HR has much to teach executives (and to deliver on) when it comes to helping organizations realize the powerful combination of human–technology collaborations for driving performance, productivity, and innovation in the future of work. For now, I'll leave smart, connected HR strategy as a topic for another book.

To help raise further awareness and understanding about the case for deep development, I have summarized two additional studies that show how deep development will create strategic value in the future of work. These data are offered to help you educate and counter the mindset that technology will be the *sole* source of value creation in the future of work.

Findings from Report #1

In conducting research for this section on the future of learning and development, I discovered a September 2016 study conducted by Deloitte titled, *Transitioning to the Future of Work and the Workplace: Embracing Digital Culture Tools and Approaches*.[3] This study anonymously surveyed 245 CEOs representing from small businesses (revenues less than $50 million) to large global businesses (revenues more than $10 billion). The study revealed important leadership lessons based on what CEOs are thinking about the future of work.

In Table 5.2, I've not only summarized Deloitte's findings, but I've also translated those findings into a set of "related skills" tied to deep leadership development.

Skills for Transitioning to the Future of Work

The Deloitte study makes a powerful conclusion about the future of work. According to the authors, "[the future of work] will require leaders to *act increasingly as network architects* and *role models* for the new ways of working" [emphasis added] (p. 10).

Based on this report's conclusion, I have two predictions: (1) Most leaders won't have the foggiest idea what "acting as a network architect" means or how to do it. This is consultant speak. And (2) even if many of today's leaders learn that future-focused leadership requires them to model effective empowerment, informed risk taking, team development, collaboration, enterprise design thinking, and so on, some simply won't be able to make the transition from old to new ways of leading. To have any chance at developing leaders for the future of work, organizations have to start redesigning their human development approach to meet the leadership demands of the future.

Redesigning how organizations *do* development is a much heavier lift than plugging in a new software system like Workday. Once HR places new technology tools, they have to learn how to use the tools, and more importantly, do something useful with the tools. The Five Principles of Deep Development discussed earlier will help leaders redesign and align the rest of their human development approach (i.e., strategy, structure, people, etc.) with their technology.

The bottom line is this: it will take more than technology to deepen development and to build the type of learning organization that organizational

Table 5.2   Skills for Transitioning to the Future of Work

Leadership Lessons	Related Skills	Why It Matters
#1 Leaders must learn to actively develop and disseminate culture.	Vision casting; role modeling; motivating and influencing large networks of people	Of CEOs surveyed, 69 percent say culture is critically important to realize their organization's mission and vision.
#2 Leaders must learn to improve communication transparency, efficiency, and effectiveness.	Leading mobile teams; collaborative decision making; engaging temporary workers	Of CEOs surveyed, only 14 percent are "completely satisfied" with their organization's ability to communicate and collaborate.
#3 Leaders must learn to engage a diverse multi-generational workforce.	Multicultural collaboration; knowledge management; coaching and development	Only a third of millennials feel their company is making the most of their skills and experience.
#4 Leaders must learn to accelerate change through collaboration.	Design thinking, team communication, advanced collaboration	Of CEOs surveyed, 57 percent believe identifying and exploiting new business opportunities is the most important outcome from learning to collaborate more effectively.
#5 Leaders will need to learn new digital tools to master communications and collaboration.	Virtual teaming, relationship management, cross-cultural team development, digital and social tools literacy	Of CEOs surveyed, 72 percent see cross-cultural and virtual teaming as becoming normative and significant in the next five years. And 65 percent believe we will use other tools more and email less to communicate within our organizations over the next five years.

scholars have described for more than two decades. True learning organizations engage in deep learning wherein employees are free to question assumptions and leaders have the fortitude to change behaviors that no longer serve the organization's purpose. There simply isn't any other way to harness technological power and human potential unless organizations get better at deep development. It cannot be overstated: leaders play a critical role in building deep development systems for meeting future organizational demands.

Findings from Report #2

My research led me to another report that supports the importance of development for building the organization of the future. This report was published by the Boston Consulting Group (BCG) in March 2017 and is titled, *Twelve Forces That Will Radically Change How Organizations Work.*[4] BCG's report analyzes 12 technological, market, and social forces that are driving demand for talent and development at the individual, team, and enterprise levels.

The forces impacting the **demand for talent** should come as no surprise to you, given what I've presented in this book so far. These forces relate to digital productivity and technology, and include *automation, big data and analytics,* and *access to information and ideas.* Responsive 4IR leadership, by definition, means that leaders will value and embrace these mega trends as both an opportunity for innovation, and a legitimate talent and organizational development risk that must be monitored, managed, and mitigated.

Responsive leaders and organizations are actively seeking new and different types of people to staff new and different roles, lead new functions, and bring new ideas to life. The problem, however, is that recruiting alone isn't enough. There simply aren't enough people with the right talents to "buy." Therefore, organizations must reevaluate their talent development strategy, asking themselves, Is our development strategy deep enough?

Demand for digitally savvy, business-minded, creative problem solvers— like Nihil—is at an all-time high. And, in some specialized areas, demand for talent is far outpacing supply. This is great news! The robots aren't winning in the areas of smart, creative, and empathic work, and won't for quite some time, despite all the fear mongering that some so-called futurists and sci-fi fanatics are propagating.

But, if leaders want to win the war for smart, creative talent, they have to get better at attracting AND developing top talent across their careers. Millennial employees and leaders with the right skills can, will, and already are demanding more from their employers. Leaders are hence faced with a triple threat: higher demands from customers and shareholders for innovation, higher demands from employees for a great culture and lifelong learning, and greater competition from competitors who are actively pursuing growth and development strategies of their own.

A smart, connected workforce is full of "always-on" learners who want to maximize their development and fully live out their purpose and passion. Smart, connected employees want a strong culture, purpose, and well-being. Deloitte's millennial survey shows that 77 percent of millennials are involved in a charity or "good cause," 76 percent believe that business should be a positive force for social impact, and 88 percent believe business is a force for social change.[5]

However, despite what the workforce of the future wants, organizations aren't necessarily delivering it. According to Gallup's millennial research, only 40 percent of millennials feel connected to their organization's mission.[6] Essentially, companies are losing the "culture battle" that they've only begun to take seriously over the last decade. If your company or organization has just started to build a data-driven approach to culture development, then you're likely behind the times. Deep culture change takes three to five years and usually requires a redefinition of your leadership and organizational development strategy.

The BCG report goes on to identify six additional factors that are impacting the **supply of talent** for meeting the increased demand for new and different kinds of work. The first, as mentioned previously, is *skill imbalances*. There is a rapid evolution in the skills required to meet the creative and technological demands of 4IR jobs. In fact, a recent Gartner study predicted that globally 1.8 million cybersecurity jobs will go unfilled by 2022.[7] And, according to a project of the Stanford University Program, more than 209,000 cybersecurity jobs in the United States are unfilled, and postings are up 74 percent over the past five years. Similarly, the demand for information security professionals is expected to grow by 53 percent through 2018. These statistics are the result of skills imbalances that will take time to balance through learning and development.

The second factor BCG identified that's impacting talent supply is geopolitical economic power and mobility. Work is more mobile than ever before, yet political, economic, and policy-driven barriers exist, blocking the free flow of talent across organizational and geographical borders. The BCG study estimates that there are more than 60 "digital hot spots" across the globe, where entrepreneurialism, education, and enterprise are present and highly integrated. This report estimates that 90 percent of these digital hot spots are in the United States. This means that leaders are going to have to become more competitive and more attractive for recruiting top talent in diverse areas, while also growing and attracting talent from other areas to feed the talent demands of the future. It also means that leaders will have to learn how to forge public–private communities of experience and practice to tackle issues like immigration policy to fuel growth and innovation.

Leaders must learn and develop a more global mindset and workplace culture so that they can meet the workforce demands of the future. Martin Ford has written a tremendous book called, *Rise of the Robots*.[8] Ford makes the following argument in favor of public–private collaborations on the future of work relative to global inclusion. Ford's analysis shows the combined population of India and China is roughly 2.6 billion people or more than eight times the population of the United States. If we were to just look at the top 5 percent of these populations in terms of cognitive ability, that's about 130 million people or over 40 percent of the entire U.S. population. Ford's conclusion: there are more smart people in India and China than there are people in the United States. Consequently, global inclusion can help solve the skills imbalance problem for the workforce of the future. This will require greater career mobility, workplace flexibility and inclusion, and smarter immigration policies.

The third factor impacting talent supply is changing workplace culture and values among talented people and future leaders. Employees are starting to demand workplaces that have an increased focus on well-being, diversity and inclusion, individualism, and entrepreneurialism. The BCG report goes into depth about each of these forces. The bottom line is this: organizations who are able to woo top talent with their "cultural value proposition," be it "purpose," "perks," or "performance," will fare far better when it comes to filling future jobs. Smart, connected, always-on learners want flexibility, meaningful work, leaders they believe in, and fair pay. So many companies screw this up when the solution is neither a secret, nor difficult to implement.

Finally, the BCG report addresses the serious demographic shift occurring around the aging workforce. I've saved this one for last because most people are aware that the global population is getting older—and that this "graying" population is a huge percentage of the workforce. But here's something you might not know: by 2035, one in five (that's right 20 percent) of people worldwide will be 65 or older. That's a lot of old people! In fact, over the last four years of traveling through Midway International Airport on a weekly basis, I've observed the number of "pre-boarders" with wheelchairs and walkers increase during the Southwest boarding process. Anecdotal data? Sure, but a sign of the future of travel!

Based on a number of simulations, the BCG report predicts a global labor crisis in the next 15 years. The crisis will impact the 15 largest economies globally, including three of the four BRIC (Brazil, Russia, India, and China) economies. BCG's estimates show that these 15 economies make up 70 percent of global GDP. Therefore, the impending global workforce crisis will impact almost every major multinational company. If that isn't a "burning platform" for investing in developing talent and creating a compelling cultural value proposition, I don't know what is.

From Raising Awareness to Designing a Deep Development Approach

I want to be clear about leadership's role in fixing the lag in learning and development. To be successful in the future, leaders don't have to become experts in adult learning and deep development. However, smart, connected leaders must understand the essential role that learning and development play in driving responsible and sustainable organizational outcomes. Fourth industrial revolution leaders must also value learning and development in ways that leaders of previous industrial eras did not have to. This is because the work of 4IR is more complex, volatile, uncertain, and ambiguous.

Although the previous section focused broadly on the data in support of a deep development strategy so that you can start raising awareness around this important issue, this section offers a road map for building a deep development that produces results. I've organized the road map to deep development into a series of tactical steps. These five steps needn't be followed sequentially. My recommendation would be to choose one or two of the steps, and start improving learning in your organization today!

Improving development not only helps future-proof your workforce, it will also help curb wasteful spending. Every year companies spend billions on training and development interventions that don't work. The exact cost of training is hard to estimate when you factor in all the direct costs (i.e., cost of training materials, staff, etc.) and indirect costs (e.g., nonproductive time, etc.).

According to the *2017 State of the Industry* report by the Association for Talent Development (ATD), organizations spent a whopping $1,273 (USD) per employee on direct learning and development activities in 2016.[9] This number was up just slightly from the 2016 report, which showed organizations spent $1,252 per employee. For a company of 1,000 employees, that's a learning budget of $1.25 million. A company with 100,000 employees would spend an estimated $127.3 million on learning and development. Using this cost per employee, roughly how much would you estimate that your organization spent on training that didn't work?

Bersin estimates that U.S. organizations spend about $70 billion per year on learning and development. Bersin estimates that globally, organizations spend about $130 billion. This is consistent with *Training* magazine's survey of more than 777 respondents from small, medium, and large companies, which found that training expenditures in the United States were around $70.6 billion.

According to *Training* magazine's 2015 annual report,[10] 28 percent of direct learning expenditures were outsourced or went to external activities. In addition, employees average about 33.5 hours of training per year. Although these reports are based on training and development professionals' self-reported data, assuming they are somewhat representative, that's a lot of time and money spent on learning! Let's hope that those organizations are getting their RODI and not just piddling away $70 billion on shallow development that doesn't add value.

What's more shocking is that despite the huge spend on training, leaders require very little evidence of impact or RODI. Smart, connected workplaces have a learning and development strategy. These organizations differentiate between *required* development (e.g., compliance training and credentialing), *career growth* development (e.g., horizontal skills development and some vertical leadership development), and *reward* development (e.g., sponsoring a well-being initiative or passion project of a team of employees). Finally, deep development strategies and systems have quantitative and qualitative analytics in place to document RODI and business impact. Again, you don't have to eat the whole elephant in one bite here, but once you understand the business case for deep development (and for curbing spending on shallow development), you have an obligation and financial responsibility to start designing a deep development approach that gets results.

Deep Development Strategy Checklist

This checklist is more than a series of questions. It reflects a thought process and mindset that smart, connected leaders use to think about deep development and organizational performance. If you answered "No" to any of these questions, then there is an opportunity to deepen development strategy.

Step 1: Build a Deep Development Strategy

Many companies don't have a deep development infrastructure because they don't have a strategy for building a smart, connected workforce. Your deep development strategy is an essential element of your overall talent strategy, and it is essential for building an effective development system. Use the checklist in Table 5.3 to assess your development strategy.

Step 2: Develop a Plan to Implement Your Deep Development Strategy

Your development strategy must have an accountable executive who owns the strategy and the plan. If you don't have a chief learning officer, then appoint one. If you have a chief learning officer and you don't have a development strategy or plan, then put her or him on notice. Development is too important to the future of your organization to delay or tolerate poor performance.

Generally speaking, your development strategic plan should be an integrated plan that includes the following elements: program management, curriculum design, development and monitoring, content curation, change management, stakeholder management, communication, leadership development, management development, team and organizational development, skill developments, technical/professional certification/continuous education, IT, communication, and analytics.

Although this is not an exhaustive list of planning elements, it's a very good start. One of the most important factors in designing a development strategic plan is input from business and/or functional leaders. Leaders must feel shared ownership for learning and development among their workforce. Because let's face it, leaders' ability to execute on organizational and/or business strategy depends on their employees' capabilities and "always-on" learning.

Table 5.3   Tool #19: Deep Development Strategy Checklist

#	ITEMS	YES	NO
1	Do we have a business strategy?		
2	Do we have a development strategy that aligns with our business strategy?		
	If you answered "No" to #2, start with a deep development strategic planning session. Include accountable executives for driving the business strategy. If you answered "Yes," use the rest of the items on this checklist to assess the strength of your development strategy.		
3	Does our development strategy have a three- to five-year time horizon?		
4	Does our development strategy identify measurable success factors?		
5	Does our development strategy have an analytics plan?		
6	Do we have measures to determine how consistently we're delivering on our promise to employees and customers?		
7	Does our development strategy have a budget funded through the business units that it supports? In other words, do different departments have "skin in the game"?		
8	Does our development strategy balance on-the-job, self-paced, peer-to-peer, and leader-led learning modalities?		
9	Does our development strategy scale to individuals, teams, and the organization?		
10	Does our development strategy rely on business leaders to help implement it (i.e., are leaders teaching and coaching)?		
11	Does our development strategy empower employees to design their own learning and development agenda?		
12	Do we report out on the impact of our development strategy on a quarterly or annual basis?		
13	Does our organization recognize and reward developmental milestones and achievements?		
14	Does our development strategy support vertical development (e.g., mindset change)?		
15	Do we leverage our development strategy to recruit and retain our superhero talent?		

Another important element of the plan is feedback. The communications and analytics around your development strategy should play a central role in testing, refining, and evolving your development strategy as organizational needs change. You'd be surprised how many learning and development leaders don't even know how many courses (e-courses or in-classroom) they offer employees. They pay outrageous fees for content that was curated years ago, and it basically sits on an outdated learning management system that few people use. Which brings us to Step 3.

Step 3: Conduct a Deep Development Inventory and Audit

Because most organizations lack a future-focused development strategy, most clients don't know what they *have* and what they *need* when it comes to their learning and development plan. Once you have a development strategy, one of your first steps in your plan should be to inventory what's in your learning and development portfolio. When I was a young organizational development consultant, one of my mentors advised me: "When in doubt, count." If you don't know where to start when overhauling your learning and development infrastructure, just start counting!

Here's a list of 10 things that you should inventory when conducting a deep development audit.

Tool #20: Ten Items to Inventory during a Deep Development Audit

1. The number of **employees** that you have
2. The number of employees **at each level** (individual contributor, team lead, senior leadership)
3. The number of **learning offerings** that exist for individuals versus teams; individual contributors versus team leads; managers versus leaders; and technical skills versus soft skills, and so on
4. The average number of **dollars spent annually** on learning and development total per employee, leader, manager, individual contributor, and so on
5. The number of **ways in which content is taught** (web, in-person, blended learning, self-paced, instructor-led, etc.)
6. The number of **vendors** you use to curate content and what their philosophy is on learning and development (some contradict each other, and this will confuse the heck out of your employees)
7. The number of **dollars** you spend on external learning and development vendors
8. The number of **hours** employees spend in formal learning and development, the average cost of learning per employee, per hour on an annual basis

9. The **impact of learning** and development on business outcomes
10. Overall satisfaction with learning and development among business leaders and employees. Use a simple pulse survey, such as, "On a scale of 1 to 10 (1 = not at all satisfied; 10=extremely satisfied), how satisfied are you with our company's training and development offerings?"

These items are, of course, the bare essentials of a development inventory or audit. But if you have data on these 10 things, you'll have a pretty good idea of where to begin to refine your development plan for the next few years.

Once you've got the basics down, you can begin to inventory advanced aspects of your development portfolio, such as:

- How well aligned is our portfolio with best practices in adult learning?
- How well integrated is our development strategy with performance development programs?
- To what extent does our portfolio develop employees vertically? To what extent does our development approach develop employees horizontally?
- What do our highest performers know, and what are they able to do that sets them apart from the rest?
- What's our return on development investment?

Don't be intimidated if you don't know the answers to these questions. Most leaders cannot even answer the basic items listed. The point of auditing and inventorying is to simply get started and learn more about what the current state of your development system.

Step 4: Starting Small Is Better Than Not Starting at All

Leaders don't have to wait for a full-blown learning and development strategy to make a difference in developing people in their organization. Starting with small changes is better than not starting at all. Gallup research shows that 93 percent of millennials *left a company* when they made their last job role change. Among the many reasons why a millennial might feel compelled to leave an organization for a job change, a perceived lack of learning and growth opportunities is a likely culprit. So, even a simple conversation with more senior leaders about an experience that they're having can make a big difference.

Team leaders should be proactive in having career growth and development conversations—especially with their "A players." Many team leaders lack the awareness, skills, and courage to have candid conversations with their followers about how engaged they are with their work or what they see for themselves as their next career move. Conversations about career growth and aspirations can often feel awkward for more junior employees due to a lack of transparency, trust, or suspicion. Developmental conversations can also feel awkward for leaders because they don't have a framework or milestones for gauging progress (e.g., around key developmental experiences) and giving useful feedback.

In the past several years, organizations I consult with have been moving toward more engaged forms of experiential learning and development. Experiential leadership development models target specific "breakthrough experiences" that lead to rapid learning, growth, and development for tomorrow's leaders. The analysis and insights that come from a breakthrough experience system help organizations target on-the-job development for next-in-line leaders—not with courses and web-based modules, but with real-world stretch assignments like owning a budget or profit and loss statement, leading a transformation, or facilitating cross-functional collaboration. The learning paths for each breakthrough experience can be configured to meet the strategic needs of the organization, and the individual development needs of high-potential leaders. High-potential supervisors can be trained on how to interpret and use breakthrough experience analytics, to coach high-potentials before, during, and after their key experience, and to measure success over time.

The breakthrough experience process and analytics provide a structure and rigorous measurement around on-the-job development, helping organizations solve several common development problems. First, key experience analytics focus and improve the objectivity of talent reviews. It takes the guesswork and some of the emotion out of debates surrounding a person's experiential maturity or "readiness" for taking on a more challenging role. Second, the data and advice that result from a breakthrough experience analysis improve the analytical rigor around succession planning and scenario planning during times of uncertainty and change. Finally, the developmental tools make it easy for leaders to "own" and support the development of other leaders. As one client said, "We use a 70 [job-related experiences], 20 [interactions], 10 [formal training] rule in our development plans, and I never know how to advise my direct reports on the 70 percent. This provides a clear road map and coaching guide for our interactions. That's 90 percent of the work right there!" And it isn't just me who is observing a movement toward on-the-job development. According to Bersin's *2017 Human Capital Trends* report, companies with strong experiential programs rose from 47 percent in 2015 to 64 percent in 2017. That's a massive shift in experienced-focused, on-the-job development, and further evidence that organizations are taking deep development more seriously. As my previous client said, "Iron sharpens iron," and breakthrough experiences provide a useful framework for ensuring the right process and outcomes for deep development of tomorrow's leaders.

Table 5.4   Tool #21: Career Growth and Development Conversation Guide

Item	Question	Notes/ Observations
1.	What attracted you to this field/job?	
2.	What has been the most impactful experience that you've had in your career?	
3.	What are some insights or lessons you've learned from these experiences?	
4.	Are you aware of your future career path at [our organization]?	
5.	What are your future career aspirations? How can I help you achieve these goals?	

To help you start to improve the quality of your career growth conversations with other leaders, I've included a Career Growth and Development Conversation Guide (Tool #21) for you to use with your superstar employees or peers. This, of course, is not a full-blown research-based conversation about the breakthrough experiences that matter most in your organization, but it can help you make a quick start in fostering deeper and more relevant conversations about development. Try using it today. Express genuine curiosity about a person with whom you're talking, and you'll be amazed at what you learn. If it works, send a copy of the exercise (or better yet, the book) to a colleague.

Career growth and development conversations are designed to help leaders learn more about what others care about. They can also be used to better understand career awareness, aspirations, areas of strength, and areas where people want/need help.

As you can see from these questions, career growth conversations can be relaxed and very revealing if you approach them with curiosity. These conversations require sincere care and trust for the people with whom you're talking. Career growth conversations can be uncomfortable when the people involved lack trust and transparency. However, they can also be transformational moments that improve trust, collegiality, and even friendships between leaders within an organization, improving collaboration and decision making down the road. You may have noticed that the questions are all openended and designed to spark reflection, self-awareness, and self-discovery. Moreover, the questions, if asked with genuine curiosity, show that you, as a leader, care about this person's growth and development. This exercise can create a small shift making a big difference in someone's personal and professional life. I hope that you try this exercise and reflect on the difference that you made in another person's career.

Step 5: Measure for Success

How does one determine whether learning and development are happening in their organization? Do you know what the specific barriers to development are in your organization? Do you have a sense of how much the pace of learning in your organization is impacting innovation and/or business results? These are really important questions to answer in the new reality of work.

Measuring development can occur at the individual, team, or organizational levels. There are two primary types of measurement that matter in the learning and development domain: *assessment* and *evaluation*. Assessment uses methods that are designed to uncover opportunities and bring a sense of where your organization is at relative to other parts of your organization or industry. Assessment measures use "quick and dirty" methods to "ballpark" the current state of development among teams, or the entire company, and addresses topics like knowledge, skills, experience, performance, and so on. Assessments can include objective measures (e.g., surveys) or subjective assessments (e.g., verbal feedback and employee anecdotes). Leaders should use focused assessment methods to understand the current state of development among individuals, teams, and their company.

The second type of measurement that can be used to deepen development in the workplace of the future is evaluation. Evaluations are simply a determination of efficiency or effectiveness. Leaders use evaluation metrics to determine how well learning and development are occurring at the individual, workgroup, or organizational levels. Evaluation metrics could include things like how much people know, how many employees have learned a particular skill (e.g., Python programming language), how many courses or coaching sessions were conducted in a given year, or what the average level of performance within a given job, business, or geographical region is.

In the Table 5.5 Tool #22, I've included a simple table to provide you with a way of visualizing the two primary types of measurement that matter for deepening your development strategy. This table also includes some sample questions that leaders should be asking within each of these domains.

Assessments and evaluations of efficiency and effectiveness are the most common learning and development measures. Although these may seem like simple measures, many clients with whom I work struggle to answer these simple questions. The most mature learning organizations are using more advanced metrics and analytics to predict performance and/or to target development to improve effectiveness. Advanced development analytics are designed to be "leading" indicators of performance. These advanced analytics help leaders predict patterns in employee behavior and performance. Using different statistical models and parameter estimates, future-focused leaders can now

Table 5.5   Tool #22: Deep Development Success Measures

	Efficiency	Effectiveness
Assessment	• Do managers feel employees have the right information and skills? • What development opportunities do employees want? Are these aligned with our organizational strategy?	• Do employees who've participated in specific types of learning activities demonstrate better performance than employees who did not partake in the activities? • What do employees think about the relevance of our development offerings for improving their performance or advancing their career?
Evaluation	• How many courses have individual employees taken in the last year? • What learning and development topics are of increasing interest and/or decreasing interest?	• What measurable improvement can be calculated as a result of learning and development investments? • Are employees or teams demonstrating a measurable improvement in performance? Retention, etc.?

leverage predictive analytics to ensure that managers will succeed in their roles before they are hired or promoted, identify turnover risks among highly valued employees, and improve safety and quality.

These kinds of predictive analytics are typical of the smartest of the smart, connected organizations. The insights and decision-making power of predictive workforce analytics are extremely valuable. They contribute to deep organizational learning, enterprise development, and performance. As 4IR technologies evolve and make their way into the world of digital human resources, deep development analytics will become even more commonplace. However, at present, most organizations lack the sophistication, infrastructure, and HR capabilities to know what they should be measuring, how they should be measuring it, and what to do with the data once they have it. This will change as talent and resources become even scarcer. Leaders will have to demand more from their people analytics function and deep development at all levels of their organizations.

Chapter Summary

1. Workplace learning and development are rapidly changing. Development is no longer a "nice to have"; it's a must-have for fueling 4IR survival and innovation.

2. Smart, connected leaders must build a deep development infrastructure and hire "always-on" learners who are hungry to learn throughout their careers.

3. Creating a dynamic development ecosystem is essential for attracting, enabling, and retaining the workforce of the future. This requires a deep development strategy, processes, and technology.

4. In addition to introducing key findings from several important industry reports, this chapter introduced five tools for improving development in your organization at the individual, work group, and enterprise level: the Five Principles of Deep Development, the Deep Development Strategy Checklist, Ten Items to Inventory during a Deep Development Audit, the Career Growth and Development Conversation Guide, and the Deep Development Success Measures matrix. Finally, this chapter has summarized several findings from industry reports that will help you make the case for, or start conversations in your organization about, the need to deepen development to get your staff or workforce ready for the future of work.

Recommended Actions

1. Conduct a Deep Development Audit and inventory the 10 things that matter.

2. Assess and evaluate how efficient and effective learning and development are in your organization. Where is the most learning happening? How can you leverage this?

3. Finally, have a career growth and development conversation with a direct report and a peer.

CHAPTER SIX

Discernment

What advice should we give the virtual Howard Schultz in Chapter One? If you recall, he was faced with a choice between a massive layoff of his workforce of 254,000 employees and a strategic growth opportunity to provide highly differentiated mixed-reality experiences for Starbucks' customers.

How would you coach Gerry, the SVP in Chapter Two, who was working 18-hour days at the expense of her presence of thinking, feeling, and action? How can Gerry gain presence of mind to more seamlessly manage her energy between work and family commitments? As a shareholder of her company, how much would you estimate that Gerry's lack of presence, and that of the employees for whom she is a role model, is costing the company in the form of burnout, turnover, health care costs, absenteeism, and lost productivity?

How should consultants have been advising Jane and Acme's board of directors from Chapter Three about creating a future-ready workforce and positioning their company for success in the digital age? How much longer should companies wait before digital disruption leaves them in the dust?

And what about leaders like Dimitri in Chapter Four, who are struggling to navigate the tensions and power relationships in highly collaborative organizations? How can we help leaders in his position choose different ways of collaborating and communicating so that they can more fully (and more quickly) leverage new business strategies and organizational structures?

Finally, what steps would you advise Sarah to take to help leaders in her company better develop smart, connected employees like Nihil? How can leaders in organizations create a culture of deep development to attract and retain the best talent? And, more to the point, how can someone like Sarah—who doesn't have a clue what a smart, connected employee like Nihil knows or is capable of doing—support the learning, growth, and career development of the people they lead?

These are complex questions that define leadership in the digital age. Digital and other fourth industrial revolution technologies will significantly impact how successful leaders create shared futures with their followers and peers. At the core of these digital disruptions, for leaders, is a fundamental demand to be a good human being. Being a good human leader means, being *present*, mastering *velocity* and *collaboration*, and *developing* a smart, connected workforce that is always growing through deep development.

For the nine months I researched and wrote this book, every day seemed to produce new evidence that the pace of 4IR change is increasing. Each day the headlines also produce new case studies that these fundamentally human demands of leading in the digital age are becoming more relevant. The case studies in the news also bring up a fifth and final demand of leading a future-ready organization, and that's simply doing what is right and using sound ethical judgment. As the cases in this book have shown, ethical judgment isn't always clear and isn't always the path of least resistance.

The mega trends that define the digital age have the promise of leading to new jobs, educational opportunities, and technological breakthroughs. And these trends also have the potential to threaten social stability and confidence in social institutions. Leaders cannot afford to act as passive spectators in the future of work. They must exercise ethical influence and agency in shaping the future of work and the future of life. The types of leadership decisions and choices around fourth industrial revolution challenges demand what I call leadership discernment. *Discernment* literally refers to the capacity to distinguish, decide, discriminate between alternative options, and to judge well.

I have chosen the word *discernment* over *decision making* because I think the word *discernment* carries a bit more weight. Everyone *makes* hundreds of decisions per day: what to wear, what meetings to attend, how to react to bad news, how to celebrate good news, and so on. Decisions mostly occur when we are on autopilot.

Discernment, on the other hand, is a more thoughtful, deliberative, and intentional process of judgment and ethical choice. Discernment takes care, time, and effort, especially when it involves multiple stakeholders. Discernment requires an elevated form of dialogue and communication between different parties to create shared understanding in one another's purpose, values, and goals, and to determine how to build a shared future together.

Because the fourth industrial revolution is ripe with possibilities and pitfalls, I believe the types of choices leaders are being called to make (and will continue to be called to make) require a more intentional approach for making ethical decisions together. My advice to leaders faced with challenging 4IR choices is this: Default to *decision* for the small stuff. Discern over the important stuff. And try to have the presence of mind to know the difference.

What Is Discernment?

Discernment is derived from the Latin word *discernere,* "to sift or distinguish." In the context of complex 4IR leadership decisions, discernment is concerned with the very applied and practical challenges of helping leaders sift through all issues to figure out what their organization *ought* to do. My goal in this ethics chapter is not to prescribe standards of what is right or wrong, but to help leaders see that they have some really important choices ahead of them. I also want to equip leaders with a new process for discerning the best ways forward for their organization, and for improving the quality of decisions they make for their organization.

The discernment process is an essential element of a future-focused organizational strategy. However, instead of speculating around what *will* happen in the future (e.g., around robots stealing all the jobs, etc.), this future-proofing strategy demands that leaders ask, "What *should* happen in future?" This mindset embraces "the ought to do" ethical mindset that responsible fourth industrial revolution leadership requires. Asking what the future of work *should* consist of places ethical agency in the hands of humans who have empathy, values, beliefs, and compassion. Paradoxically, these are uniquely human capacities for moving boldly into the uncertain digital future. The capacity to discern what is good, right, just, and fair is one of the greatest leadership tools that differentiates the best leaders in history.

Conversations about what an organization ought to do to prepare for the future aren't easy. I know, I facilitate lots of these conversations among senior leadership teams. Leaders such as you know that making decisions that have significant ethical consequences are hard on your own. In fact, many leaders try avoiding these conversations altogether—these folks don't last very long in their positions.

Future-focused discernment and decision making require presence, collaboration, and a well-developed communication tool kit. I've observed leaders and leadership teams struggle with making and implementing even the simplest of decisions, such as which employee should get a small cash award for exemplary performance? These same leaders struggle even more when you add complexity, ethics, and values to the conversation. One of my litmus tests for determining whether or not leaders may struggle with future-focused decisions is simply to ask, "How are important decisions made on your team?" Their answer tells me about their decision-making process and their mindset about risk, inclusion, collaboration, quality, and so on.

Discernment principles and practices help focus and improve decision making. As I'll show you in the following, the discernment process looks and feels a bit different than the typical decision-making process. Discernment processes are explicit and unapologetic about the purpose, mission, vision, and most importantly, *values* that inform how an organization moves into the future. Right from the start of a discernment process, leaders must ask, How will this impact our organization's purpose and mission? The answer to this question determines whether a formal discernment process is warranted. Discernment is also more deliberate about what voices are included. An inclusive organizational culture is concerned with more than just including diverse people. Inclusive culture includes diverse points of view, values, and beliefs about how the organization should move into the future. Finally, discerning organizations and leaders use their values and organizational beliefs as criteria and the basis for evaluating decision quality, learning, and improving decisions over time.

Some organizations are more mission-centered, which means that values-based deliberation might seem a more natural fit. These "usual suspects" include faith-based organizations, education, government, nonprofit organizations, and social justice organizations. However, I would argue that making socially responsible decisions is of increasing importance in private, for-profit, secular organizations as well.

Throughout the second and third industrial revolutions, corporations and for-profits have been primarily concerned with *efficiency* and *effectiveness* in the name of growth and profit. Sure, there have been many exceptions, but for the sake of argument, this has largely been the case. The fourth industrial revolution, however, is starting to challenge these types of organizations to get better at engaging questions of ethics and doing what is right.

For those of us who have been around for a while, corporate social responsibility initiatives are a relatively new phenomenon. This is, in part, required of corporations because of increased transparency, public demand for accountability via social media, and the need to protect brand reputation. But socially responsible companies like TOMS Shoes, which donates a pair of shoes to a person in need for every pair that's purchased, have differentiated themselves by doing what they believe is right. We could say that "doing good" is both a requirement for 4IR organizations and, perhaps, a source of strategic advantage. It's hip to be do-gooders, and, as it turns out, there's a whole customer base that prefers to support good companies.

Having said that, meeting the demands of shareholders, customers, and the public interest is a delicate balance. An organization's ability to balance purpose and productivity is hugely influenced by the quality of the decision-making processes used. I call this the triple aim of discernment, that is, balancing questions of *efficiency*, *effectiveness*, and *ethics*. I've summarized these questions of the Table 6.1 on the triple aim.

Triple Aim of Fourth Industrial Revolution Discernment

If an organization's ability to balance purpose and productivity is determined by the decisions it makes around the triple aim, then doesn't it make sense to build capacity around discernment and the triple aim? Doesn't it make sense

Table 6.1   **Triple Aim of Fourth Industrial Revolution Discernment**

Efficiency	Effectiveness	Ethics
• How do we reduce cost? • How do we maximize our resources? • How can we operate in a leaner fashion?	• How do we grow revenue? • How can we innovate and differentiate ourselves among competitors? • What investments in innovation will yield the highest returns?	• What should we do to produce the greatest good? • How do we do the least amount of harm? • How can we balance purpose, profitability, and prosperity for all?

to give leaders some choices and tools for making different kinds of decisions under different circumstances (e.g., those that threaten fundamental values)? One doesn't have to look very far to find examples of corporate fraud, government scandal, and bad decision making to make a compelling case that leaders need better decision-making and discernment processes and skills.

Toward this end, the remainder of this chapter will focus on the fundamentals of the discernment process. Specifically, I'll contrast the steps in a simple decision-making model with those of a discernment model. In addition, I will provide advice for building discernment capability among team and organization leaders. Finally, I will offer some research-based recommendations for improving the efficiency and effectiveness of decision making and discernment in your organization.

Two Models: Simple Decision Making versus Discernment

Models are useful tools in that they are abstract representations of what they intend to represent. Take, for example, an old-fashioned model car. It's not a real car with all the working parts and details; however, the model is a simplified version of how a particular car—let's say a Ford Mustang—is designed. A simplified model allows a Mustang enthusiast to quickly and inexpensively compare multiple versions of the Ford Mustang and "play around" with them.

If, for example, you had two Mustang models next to each other—say, a 1967 model and a 2018 model—you could compare and contrast the similarities and differences between the two models. You might ask yourself, what elements of the body design have *stayed the same* over the years? What elements of the body design have *changed*? What *purpose* do these changes serve? And you could also speculate as to *why certain features* changed?

This elementary example helps to illustrate the value of putting two models side by side. Models allow us to compare and contrast two things. They allow us to objectively examine the design of a particular thing, and they allow us to discuss and question the assumptions behind the model(s). In this 4IR leadership case, we're comparing two decision-making models: simple decision making and discernment. Leaders need to know both models because not all decisions are simple, and, likewise, not all decisions require discernment. Understanding both models will help 4IR leaders more efficiently and effectively know when to use which.

Discernment and decision making are both complex information processes that can be easily modeled. Like the Ford Mustang, decision making and discernment models are simplified versions of how the real process plays itself out in organizations. In other words, these information-processing models are not the "real thing," with all the nuances and moving parts. These models allow us to compare the costs and benefits of each. What's more, they allow us to succinctly teach others about making good, purpose-driven decisions amid complexity and uncertainty. Let's start with the simple model first.

In the Tool #23, I've listed the steps in a simple decision-making model and provided a brief case study to illustrate how this model might play itself out.

Tool #23: A Simple Decision-Making Model

In a simple decision-making model, there are roughly seven steps:

1. Define the decision to be made.
2. Gather relevant facts, data, information, perspectives, and so on.
3. Identify all the alternatives.
4. Weigh the evidence and prioritize the alternatives.
5. Choose the best alternative.
6. Take action to implement decision.
7. Evaluate the quality and impact of the decision (repeat if necessary).

Individuals or groups can use these seven steps to make decisions. In the spirit of simplicity, let's start with an independent decision. Remember Dimitri from the chapter on collaboration? At one point, he was considering going back to a sales role with one of his current company's competitors.

If you were Dimitri, here's how you might use the simple decision-making model to choose your next career move. Step 1 would entail defining the decision to be made (i.e., *To leave or stay at my current company?*). Step 2 would be to gather some information to help inform the choice and course of action. At this point, you would assess the strength of the job market, search for opportunities with other companies, or maybe you might evaluate the prospect of starting your own business. During this information-gathering phase, you might also ask what your partner or spouse thinks about this career change, or you might seek advice from a mentor, friend, or executive adviser. This step can take a while because there is a lot of data to gather and a lot of different sources to go to for information.

After gathering all the information and data, Step 3 entails identifying all of the relevant alternative options (i.e., *stay at current employer, submit a resume/ CV to competitors A, B, & C, work with a recruiter, or start a small business of your own*). These alternatives will probably require more information gathering and a brief trip back through Step 2.

Once you've identified and gathered all the relevant information about your alternatives, Step 4 involves weighing the evidence and prioritizing alternatives. You might create a "pro/con" list to compare the benefits and costs associated with staying at your current job, applying to new jobs, working with a recruiter, or starting your own firm. You might rank these or apply some other algorithm to help you prioritize the best alternative(s) courses of action.

Once you've weighed your options, Step 5 involves choosing the *best* alternative. Let's say the best alternative for you—based on your purpose, values, mindset, and so on—is to stay at your current job for the next six months, and to start laying the foundation for your own business in your spare time. Having made this decision, you then have to put together a plan to implement your decision. A simple plan of action might include writing a business plan, talking to potential investors, building your product/service offerings, creating a marketing plan, and so on.

This brings you to Step 6, implementing the decision. After six months, you actually have to resign from your current position, execute your business plan, and start your business. There are lots of barriers to implementation here: fear, changes in the market, changes in your personal situation, and the like. But assuming you overcome these barriers and successfully take action, after six months to a year, you could move to Step 7: evaluating the quality and impact of your decision.

At this point, you are taking stock of how your business is operating. How has your decision to start your own business impacted your financial well-being? Your work–life balance? Your happiness? Your relationships with your children, spouse, and so on? Upon answering these questions, you can determine whether or not any future decisions need to be made around your career (e.g., to grow your business, sell it, return to your previous firm, retire, submit your resume to a competitor of your old firm, etc.).

As this brief case illustrates, simple decisions aren't really that simple. Even with an independent decision to change jobs, there is a lot of data to gather, a lot of complexity to manage, and a lot of risks and rewards to be weighed. The "simple" decision model gets even more complex when interdependent parties have to weigh the evidence and alternatives together. This can lead to disagreement over facts and quality of evidence among the decision makers involved. This is why leadership teams struggle with even the most basic decisions that impact their organization.

Let's contrast the simple decision-making model, then, using the same job change example, with the discernment model. The discernment process has a couple of important additional steps, but also shares some commonalities with the simple decision-making model.

Tool #24: A Simple Discernment Model

1. Identify THE NEED for discernment (i.e., are there significant values-based or ethical implications inherent in this decision?).
2. Define THE ISSUES and decision to be made.
3. Gather relevant facts, data, and information.
4. Engage KEY STAKEHOLDERS who may influence and/or will be impacted by the decision.
5. Weigh the evidence against THE VALUES of the organization and its mission and purpose.
6. Consider the alternatives.
7. Make a decision.
8. Implement the decision.
9. Revisit and review the quality of the decision.

The beginning of the discernment process (Step 1) requires determining the need for discernment. Because discernment is a bit more time consuming, this is an important question to begin with. So, if you were Dimitri and considering changing your job, you might ask, "Does this decision have inherent values-based implications?" Let's say the answer to this question is yes. As the sole decision maker, you determine that your values around work–life balance and the financial well-being of your family are significant values that should inform your decision. Therefore, you should proceed with the discernment process.

Step 2, then, involves not just identifying the decision to be made (i.e., to stay, leave, or build your own business), but also to identify *the issues* at play. The issues might include things like trading time with family for a sales job that requires more travel but may yield a larger income. These values-based issues (work–life balance and financial well-being) are seemingly at odds, in tension with one another, and require discernment. Part of that discernment, as with the simple decision-making model, requires gathering facts and information (Step 3). However, because this is a values-based decision, you might have to broaden your exploration of facts and information. You might read a book by Sheryl Sandberg about work–life balance in the digital age, attend an online course about finding your purpose, and so on.

This brings us to Step 4, engaging key stakeholders who might be impacted by the decision. There is a subtlety here that I don't want you to miss: this step is about more than just seeking perspectives. It's about truly *engaging* "stakeholders." Engagement means letting others into the process, letting them know that they matter, and gathering their thoughts on how they think this decision will impact them. For example, you might call a family meeting with your spouse and children, ask great questions, and really listen to their opinions. This will aid with your discernment. Moreover, it makes the process more inclusive, which opens the process up to the possibility of being altered by the people involved. This is a very dialogue-driven way of making a decision (remember the importance of dialogue from Chapter Two?).

Another form of stakeholder engagement might include self-reflection—don't forget, you're a stakeholder too! You might reflect on your values and more deeply explore your relationship with your career. Maybe you'd seek some counseling about your relationship with money and try to figure out how "never having enough" is rooted in your childhood. Maybe you'd attend a support group for workaholics or seek out a start-up incubator of like-minded entrepreneurs. If you're a religious person, you might seek counsel from a priest, rabbi, or imam. Or you might just go for a walk in the woods for some presence of mind and meditate on the decision to be made. Discernment has a deeper, self-reflective, and spiritual side to it. In fact, in Christianity, the term *discernment* refers to the process of determining God's will in one's life or in a particular situation. Despite these religious connections, the self-reflective nature of discernment is highly practical and appropriate in secular organizations. My point is that personal discernment can involve a variety of forms of personal reflection, either by yourself or in a group—this is what Step 4 is all about.

Step 5 includes weighing the evidence against your values (or your organization's values), as the case may be. In deciding to leave your job, you would consider the pros and cons of each alternative against your beliefs and values around work–life balance and financial well-being. The reflection and stakeholder engagement in the previous steps helps with this. By looking inward and talking with others, we gain clarity about where we've been, what we value in the present, and what we *ought to do* moving forward. These two steps (4 and 5) are really the essence of discernment.

For example, after weighing all the evidence, if you're Dimitri, you might determine that your love of money is not as important as your love of family. And you might decide that the best choice for you is to NOT start your own business or travel for a high-paying sales job, but to stick it out in the company that you've given 15 years of your career to, grow your skills as a leader, develop your team, and master collaboration in your organization.

From here, Steps 6 through 9 pretty much look the same as the simple decision-making model. However, in evaluating the quality of your discernment (Step 9), you would of course weigh the impact of the decision against your values, beliefs, and so on. So, as you can see, the key differences between decision making and discernment are as follows: First, determine whether a decision requires discernment. Second, engage stakeholders (including yourself). Third, weigh the evidence against your values before you make a choice. And, finally, evaluate the decision against your purpose and values.

Based on this example, do you see how discernment explicitly brings in purpose, values, and key stakeholders? Do you see how the process is more intentional and deliberate? Do you see how it requires a bit more leadership presence and dialogue, as described in Chapter Two? These extra steps aren't that much more onerous. But they are extremely important for making high-quality decisions that align with your organization's values and purpose.

Regardless of the complexity of the decision at hand or the processes by which choices are made, our values and ethical judgments are always at play. To a greater or lesser extent, your values impact the decisions you make every day. If you purchase a luxury vehicle over a standard vehicle, you have made that decision based on things you value. If you eat an apple over a cookie for an afternoon snack, you've exercised your values. The difference between a decision and discernment is in consciously choosing before acting or reacting. This is the "responsive" nature of discernment. It's not just a knee-jerk reaction; it's a fully present, mindful response.

Given the importance of decisions with which 4IR leaders are currently faced, and will continue to face, thoughtful discernment will be a key requirement. Organizations that want to make responsive and responsible decisions around 4IR opportunities and challenges must build capacity around greater discernment and decision making.

The Value of Discernment

There is great value in discernment for organizations. I want to cover these briefly for leaders who are interested in raising awareness in their organization about the value of upping their decision-making game. First, discernment helps leaders and teams make better decisions. I shared my leadership decision-making litmus test earlier (i.e., "Tell me how important decisions are made in your organization."). What I find is that most leaders and organizations don't have a formal decision-making process, and far fewer organizations have a formal discernment process in place. In other words, despite all the important decisions that leaders make every day, most lack a framework for high-quality decision making. That's scary!

The second benefit of the discernment framework is that it can help organizations mitigate risk and maximize opportunities. Discernment and dialogue create an intentional pause for leaders to really dig into issues and decisions to be made. This pause and intentionality in gathering information and perspectives helps to surface information that would have otherwise been overlooked. Remember, we're trying to avoid groupthink! In addition, by engaging stakeholders and taking a more inclusive stance in the decision-making process, discernment can save leaders from their own "blind spots."

Thirdly, discernment builds team morale and enthusiasm to support decisions. When I was an internal leadership adviser at Trinity Health, a Catholic health care organization that has a formal discernment process, we taught our leaders that some people will be *more likely* to support decisions that they help create. And other people *will ONLY* support decisions that they help create. The lesson here, from a change management perspective, is that inclusive discernment and dialogue are engaging and can accelerate change. These are processes that bring people together, foster community, and build energy for doing good, purpose-driven work. And, they don't take that much more time. They simply require a commitment from leadership. As I've observed, the more you engage in a discernment process, the better you get at it.

Finally, discernment is a powerful method for building shared futures among communities that are experiencing stress, uncertainty, and turmoil. As I write this chapter, there are plenty of divides between different groups in the world: for example, between people of different races, religions, political orientations, and genders. Some days it feels like we are on the brink of social, economic, and political disaster. Public opinion of "leadership" is at historic lows, and the problems the world faces are getting more complex. This mix of social and technological elements, in my mind, is fueling urgency for a better model of leadership decision making about how to build a better future together.

This year, the World Economic Forum's work will center on the theme of "creating a shared future in a fractured world." This is a topic any forward-thinking person would agree is well warranted, and one that is, personally, close to my heart. As I write this, global leaders are preparing to meet in Davos to discuss and debate policy, technology, and economics around creating shared futures in the fourth industrial revolution.[1] These leaders will make important points, share eye-opening data, and write lots of white papers and books together. These efforts will continue to effect change on a macro-level.

However, at the local level, you and I should not wait for these thought leaders to figure out how to proceed. You have the power and the tools to start creating a more coherent world right now. You can start your own dialogues and bring voices into those conversations that have historically been left out. You have the tools and the frameworks in your hands, right now, to engage others in peer-to-peer dialogue or a discernment process. You have the tools to start developing plans to create a shared future and future-ready organization. All that you need to get started is the desire and courage.

To help you and your organization capitalize on the value of discernment, and to start building shared futures, the next section covers some discernment essentials, and offers advice for improving the decision making in your organization and in your life.

The ABCs of Discernment

In Tool #25, I've listed a helpful way to remember the discernment process. I call this the *ABCs of Discernment*. This is a 10-step checklist that helps leaders remember the discernment process.

Tool #25: The ABCs of Discernment

1. Assess the need (What are the triggers for discernment?).
2. Be open about the issues.
3. Consider possible reactions to the issues.
4. Discuss the facts.
5. Engage key stakeholders.
6. Formulate options and actions against your values.
7. Generate alternatives.
8. Hash it out and make a decision.
9. Implement your decision.
10. Judge, evaluate, and review lessons learned.

In this sections that follow, I'll use the ABCs of Discernment to offer tips and tools for beefing up your leadership team's discernment skills.

Discernment Skill Building

1. Assess the need (What are the triggers for discernment?).

Fourth industrial revolution challenges sometimes require slowing down to determine the upstream considerations, and the downstream implications of certain decisions. A simple stoplight indicator is a useful tool for detecting triggers that require a values-based discernment process.

Possible triggers that might require a leadership team to **"stop"** business as usual and begin a discernment process include:

- Threats to values
- Threats to brand and identity
- Major mergers, acquisitions, or divestitures that pose a threat to culture

- Opening or closing major facilities
- Workforce reductions

Triggers that require leadership teams to **"slow down"** and determine if there are unforeseen upstream or downstream implications that should be reconsidered include:

- Major capital or human capital investments
- Selection of critical leadership or board positions
- Multi-stakeholder projects or programs that require external partnership, objectivity, or expertise
- Episodic events that question fundamental mission, vision, values, or operating principles
- Social responsibility initiatives and/or opportunities for giving back to communities of interest

Finally, many triggering events could be considered **business as usual**, and likely will not require a full discernment process. Leaders in the fourth industrial revolution, however, must develop a habit of asking themselves, Does this trigger require greater discernment? Such events might include:

- Changes in organizational structure
- Communicating or implementing decisions (the "how")
- Incremental changes to technologies, systems, or processes
- Changes in employee education and/or training
- Business-as-usual operational decisions *(you should question these from time to time)*

In the event that a triggering event is determined to require a full discernment process, then proceeding with Steps 2 through 10 of the *ABCs of Discernment* is highly recommended.

2. Be open about the issues.

Complex discernment processes often require process consultation or facilitation. This can be carried out through an independent office like an ethics or integrity office, or internal organizational development experts. I've also observed discernment processes being managed out of the "office of the president." The program management and governance structure through which discernment issues are vetted and discussed plays a critical role in defining key issues, convening key stakeholders, gathering facts, and managing milestones and communication.

In many cases, external facilitators are used to ensure that stakeholders are aware of the issues, and that a constructive dialogue around the facts and issues takes place. Consultants and facilitators can also be helpful in gathering data, evidence, and/or input from key stakeholders. The value of third-party partners is their outside perspective.

Outside discernment experts are in a unique position to positively impact decisions because they are more immune to organizational history, hierarchy, and politics than insiders. Moreover, a good external facilitator can more easily see risks and call out counterproductive behaviors. As my favorite philosopher, Mikhail Bakhtin writes, "In the realm of culture, outsideness is a most powerful factor in understanding. It is only in the eyes of another culture that foreign culture reveals itself fully and profoundly."[2]

To use a different analogy, a health care leader—a former surgeon—once thanked me for raising some issues about his leadership team. He said, "You did a great job on us. You know, it's really hard to perform your own heart surgery." If you feel yourself or your team closing down or ignoring certain issues that may have detrimental consequences on your decisions, seek external advice, ask for help, and hire a moderator who is well versed in the future-focused challenges that your team is discerning. Outside perspectives can encourage your team to remain open about issues, mitigating risks of groupthink around key business decisions.

3. Consider possible reactions to the issues.

Empathy is a key 4IR leadership skill. Empathy is about having a focus on someone other than yourself. We live in an "it's all about me" culture. Being digitally connected to others doesn't mean that people are more connected in terms of having an outward focus on other people or their community. In fact, people are less connected in this way than ever before. Most of the developed world has lost its sense of tribe, village, and interdependence.

Because we are less connected, one cannot help but ask, How is the lack of deep human connection going to impact future leaders' empathy and emotional intelligence? A leader's ability to take another person's perspective is extremely valuable when it comes to discernment and decision making. Perspective taking helps us anticipate and consider possible reactions to issues, and respond to ways that are helpful, results-focused, and influential.

Here are some helpful questions for becoming more empathic and other-focused, and for improving your consideration of possible reactions to issues during the decision-making and discernment process:

- How would my best friend react to this issue?
- How would an alien from another planet who was completely emotional *feel about* this issue?

- What would an alien from another planet who was completely rational (like Dr. Spock from *Star Trek*) *think about* this issue?
- Why might someone disagree with me about this?
- Who might care about this even more than me?
- Whose point of view have we forgotten or inadvertently left out? Have we considered our customers' or constituents' point of view?
- What questions or concerns might my key stakeholders have?
- What data or facts will stakeholders care most about?

Considering others' perspectives can help you become more inclusive in your discernment process. These questions might impact the type of facts you gather and the stakeholders you include in the discernment process. Developing an other-focused mindset and exercising empathy can also help you socialize and better communicate decisions when implementing decisions.

4. Discuss the facts.

Once you've considered possible reactions to decisions, it's time to gather some intelligence and discuss the facts. Ensuring that you have asked the right questions is a critical first step. Then selecting the best method for getting answers to those questions is next. Issues that require discernment also require dialogue because what some consider fact, others might consider opinion, conjecture, or a matter of interpretation.

One of my clients once said, "We are swimming in data and thirsty for meaning." This is where outside facilitators can help teams make sense of the facts and consider different perspectives. What's more, outsiders can help teams and their stakeholders avoid groupthink. Groupthink occurs when groups or teams fail to ask enough questions (e.g., What have we missed? What would those impacted by this decision have to say about this matter? Are we being true to our values and who we say we are? etc.).

5. Engage key stakeholders.

Once you've discussed the facts around an issue, convene a dialogue about the facts and issues with other stakeholders. The rule of thumb when it comes to engaging key stakeholders in dialogue is this: always think one level broader than you think you need to, and always think one level narrower than you think you need to. Remember, you are stakeholder too!

The key to stakeholder engagement is to invite all you know (or think you know) will be impacted by a decision, and a few who won't be impacted. Inviting a few people who won't be impacted by your decisions ensures that you've covered all of your bases. Again, outsiders have a most valuable perspective and often ask very good questions during a discernment process.

Table 6.2   Stakeholder Engagement Tool

Stakeholder (Individual or Group)	What do they care about?	How can they help us?	How can they derail us?	How do they want to be engaged?

Treat stakeholder engagement as a serious work stream to be actively managed. Make it someone's responsibility to manage stakeholder engagement around a discernment process. Ensure that you have a strategy for individual stakeholders and groups of stakeholders. Use what you learn from stakeholders to gather more data and communicate more deeply about the issues. Keep peeling back the layers of the onion until everyone has a deeper understanding of the issues at hand. Your stakeholder engagement plan can help you organize what key constituents care about and how you communicate with them. See Table 6.2, Stakeholder Engagement Tool.

Make sure to revisit your stakeholder engagement plan often. Keep it updated. Be inclusive. Communicate early and often with key stakeholders.

6. Formulate options and actions against your values.

Discernment processes, by their very nature, are ethical and values based. All good decisions are about creating alternative options or ways of proceeding. Leaders and leadership teams have to get creative in their decision-making processes so as to avoid risks and maximize shared future opportunities with stakeholders.

Explicitly state your values in formulating options and alternatives. What is the best outcome that you can achieve given your values-based commitments? What does this value or commitment need to be put into action? What's the best alternative for this value to be satisfied? Are there any contradictions or tensions between the options that you've come up and your core values?

These are the types of questions and critical thinking that many people avoid. These aren't easy questions to answer, but they are important to explore. Don't be lazy in formulating options. Pose hard questions and be patient with others who pose hard questions during decision making. But at the same time, don't tolerate people who simply want to block progress. Take the best of what your stakeholders have to offer, engage the issues deeply and sincerely, then formulate the best options relative to your values and keep moving forward.

7. **Generate alternatives.**

Generating alternatives requires out-of-the-box thinking, creativity, and ideation. Your values-based options will provide a nice list of alternative ways of approaching an issue. Once you have this list, go back to your facts and lessons learned from key stakeholders. Ask key stakeholders and outsiders for alternatives, "What have we forgotten?" Don't close the discernment off prematurely until you've explored one more alternative. What's an off-the-wall wacky alternative? If you could wave a magic wand, what alternative would you produce for the team that they haven't thought of? What would it take to make that alternative a reality? Remember from the collaboration chapter, groupthink occurs, and bad decisions are made when people fail to generate sufficient creative alternatives. Interrupt the discernment process with a healthy dose of creative brainstorming. Be skeptical that you have exhausted all options or thought of all the alternatives.

8. **Hash it out and make a decision.**

Hashing it out means exploring what it will take to implement a decision once it has been made. This requires open, honest, and direct peer-to-peer dialogue. This is where presence of thinking and feeling are the most important leadership skills to exercise. This is where "stuff gets real." Who will be accountable for this decision? How, when, to whom, and by whom will decisions be communicated? What expectations will you set? What support will you need to be successful? What risks are associated with the decision? What are you doing to mitigate those risks while doing the greatest good for your organization? What information will be most important to key stakeholders? What will their initial response be? How can you meet dissent and disagreement with a productive tone, tenor, and way forward?

To be candid, I've seen a lot of clients screw up the discernment process at this point because they fail to think downstream about how others will respond. My advice is to demonstrate the courage to stand behind your decisions. Don't sugarcoat your communication. Be direct and create a unified front among the accountable decision makers involved. Don't back off the decision once you've made it. Don't throw other people "under the bus" by backing down from your part in the discernment process. Have a backbone, for goodness sake. Don't make promises you cannot keep. Don't give false hopes. Meet rationality with logic and meet emotion with empathy. Stay calm and present. Be humble. Listen to others who don't agree with the decision.

9. Implement our decision.

Implementing the decision requires follow-through. It's about doing what you said you were going to do. At this point in the discernment process, you've done all your homework, and now it's time to implement the stakeholder engagement and communications plans. There's nothing more unattractive to followers than a group of leaders making a decision and then chickening out on implementing it. Show some leadership gumption and fortitude. Use the word *gumption* more—it's going out of style and we shouldn't let it!

Provide clear descriptive and informational communication about your decision (i.e., *why we made it, what it is, and how we're going to proceed*). Lead with the "why" behind the decision. Communicate the values-based discernment process to assure people that great thought and effort were put into the process. Tell the story of how your values were used to formulate alternatives and options. Explain your rationale and reasoning for sticking to your values. Telling the story of how a decision/shared future came to be can be difficult, but "walking the talk" is extremely rewarding.

If you survive in the future of leadership, you will encounter resistance to decisions you've made. Your success and longevity will be determined not by the *lack of resistance* that you encounter, but in your *response to the resistance*. Resistance is normal and makes sense. If an event triggered a discernment process, then it means it was an important issue to people. Be sincere. Listen to the resistance. Keep in mind that you've had a longer time to get comfortable with the decision than others hearing about it for the first time. You've had time to process the facts. You've had many more conversations with key stakeholders. Honor the reactions—good, bad, or in different—of those learning about a decision for the first time. Practice exercising an other-focused mindset, which means putting others first. Use your empathy and active listening skills to meet rationality with logic and emotion with empathy. Be sincere. Show that you care about the values behind and the people implicated in the decision.

10. Judge, evaluate, and review lessons learned.

After implementing a decision, you're probably going to feel "done." You've communicated, listened, and held yourself and your team accountable for living your values. That isn't easy to do. Congratulations—you're a smart, connected leader!

However, before you move on with business as usual or the next big decision, take a moment to reflect on what you learned. What lessons did the discernment process teach you about yourself and your values? What did the discernment process teach you about your organization's readiness for the future? How did the implementation go? What risks were avoided? What challenges were encountered? Were they successfully overcome? What strengths and skills did you and your team build or rely on? What would you do differently next time?

These sound like simple reflection questions, but many leaders and organizations fail to reflect on or record their lessons learned. This is a missed learning opportunity for leaders who will follow in your footsteps. Inquiring about lessons learned is a chance to reconnect with stakeholders, to learn from experiences, to ask for feedback, and to recognize people for their contributions to the discernment process. It's also a chance to reinforce that you are a leader of a team that "walks the talk" even when it isn't easy.

There you have them—the ABCs of Discernment. I hope the tools in this section are useful to you when you sense a big, hairy, values-based decision upon you. Organizational leaders face these kinds of decisions all of the time. And, given the high-velocity nature of fourth industrial revolution changes, getting better at discernment is going to become an *even more* important organizational "muscle" to build. Leaders at all levels are going to have to get better at communicating their values through their actions. Leaders are also going to have to increase the speed and agility of their decision making. I've included the following tips for speeding up the discernment process. Use them with great care.

Improving Discernment Efficiency and Effectiveness

As we've established, fourth industrial revolution transformation is defined by rapid changes in acceleration, speed, and direction. Building organizations of the future requires agility and high-velocity change leadership capabilities, and it also demands efficient, effective, and ethical discernment and decision making. To be sure, there is some tension between deciding to do things quickly (efficiency), doing them well (effectiveness), and doing the right things in the first place (ethics). Having established some leading practices in doing the right things with the ABCs of Discernment, let's discuss now how to improve discernment speed and quality.

The first step to improving the efficiency and effectiveness of discernment is to clarify your values. If you don't know what your organization's purpose is—why it exists—then that's the best place to start. As listed at the beginning of this chapter, Steps 2 through 4 in the discernment process can really bog down a group if the group isn't crystal clear about what their purpose and values are. These steps require identifying what's at issue, gathering facts, bringing the right stakeholders to the table, and weighing evidence against your organization's values. This is the heart of discernment. If you aren't clear about your values, this will be a long, painful, and ineffective process. And, ultimately, it won't produce better decisions for your business or organization.

The second step for improving discernment speed and quality is to introduce formalized decision-making processes to your team or your organization. This chapter provides a simple discernment process that consists of 10 steps, and the ABCs of Discernment, which offer some recommendations and techniques for working each of those 10 steps efficiently and effectively. Start improving discernment processes with these tools. You don't need to reinvent the wheel or go seek a new model. If you want to get fancy, you can adapt these steps, combine the activities of certain steps, and so on. But keep the core elements intact: know your values, define the issues, let your values guide identification of evidence, alternatives, and stakeholders, and, finally, be inclusive and open.

Once you have your process designed the way you want it, explain the model to other leaders in your organization or on your team. What do they think of this process? Ask them how the model jibes with your team's existing decision-making process—assuming that you have one. (*Don't worry: lots of teams don't have one.*) And ask for commitment from your team to try the discernment process out the next time a triggering event warrants it. See if it helps you arrive at better decisions or better decision implementation.

Now, to be clear, this chapter has not tackled the science of decision making, small-group communication, or designing decision-making authority into your organizational structure. Entire books on decision making already exist. Decision rights and small-group decision research can provide a wealth of resources—which don't necessarily tackle values-based discernment processes—but can certainly be used to improve the efficiency and effectiveness of your discernment process. Go find more resources on decision making if you feel you need it.

The third step for building discernment efficiency and effectiveness is to ask for help. As you build your discernment process, ask yourself at what points in the process external facilitators or advisers might be helpful. I know resources in every organization are scarce, but as the saying goes, a stitch in time saves nine. Spending a few bucks on external help can increase speed of decision making and help your organization increase speed to achieve value from the decisions it makes. External advisory support can also help leaders tackle tough issues that no one talks about because, well, because they're tough to talk about. Sometimes the costliest decisions are the ones organizations fail to make.

External facilitators and advisers can be very helpful in building discernment competence among your leadership team. The best decision experts (that is, data experts, change managers, facilitators, etc.) can also help you build a process, infrastructure, and internal capabilities to provide decision support at every level of your organization. Get help if you feel you need it.

Here's what to look for in a good decision-making or discernment adviser: (1) They should be trustworthy; (2) they should be well versed in your organization's strategy, purpose, and values; (3) because discernment and decision-making conversations can get "messy," they need to be expert facilitators and great listeners to be able to lean into the tension in the room; (4) they should have the ability to collect and help the group weigh evidence—that is, they need good business acumen and some quantitative and qualitative research expertise; and finally, (5) they need to be free from any agenda. The best discernment experts help the team make the most effective, efficient, and ethical decisions for their organization. They ask hard questions, help the team think differently, and ensure that your team stays true to its values, while balancing situational demands (e.g., cost, competition, changes, customer feedback, crisis, etc.). In sum, the best decision consultants want to help you build decision-making and discernment competence within your organization. They don't want to make decisions for you; they want you to make decisions for yourself.

Increased discernment speed and competence require inquiry and learning. The last step in the ABCs of Discernment is about inquiry and evaluation of how well your process is working. Make the time to do this. Prioritize learning and reflection on lessons learned. The discernment experience can be a very instructive one for strengthening teams and organizations. Discernment will not only make your leadership team smarter, it will also make your team more connected.

As we close, remember that your discernment process is a risk-mitigation tool. It's a culture-enhancement tool. It's a brand-building tool. It's an employee-engagement tool. It's an inclusiveness tool. And it's an unconscious leadership bias tool. Discernment is just too important for your organization to risk leaving good, purpose-driven decision making to chance. Chance and hope are not strategies for making complex decisions. Discernment will provide the ultimate test of how well your leadership is meeting the other four demands of being a smart, connected leader that we've discussed in this book (i.e., presence, agility, collaboration, and development). As you introduce discernment, you'll start to see where the cracks in your leadership foundation are. And that's okay, because you now have some excellent strategies and tools to take action to address those weaknesses in your foundation as you build the organization of the future.

If you are a board member of an organization, make discernment a future-proofing priority. Hold executive teams accountable for triggering and implementing the process, and improving efficiency, effectiveness, and ethical quality. If you are a coach, consultant, or HR professional, ask your business partners or clients to describe their team's decision-making process. Ask them how they deliberately weigh organizational decisions against their values. Finally, if you are a leader, start asking yourself if your next big decision should trigger a more intentional discernment process.

Chapter Summary

1. The digital age and fourth industrial revolution bring major challenges for leaders to make decisions around in an effective, efficient, and ethical manner. This means that leaders will have to get better at making big, important, values-based decisions.

2. *Discernment* literally refers to the capacity to distinguish, decide, and discriminate between alternative options, and to judge well. Discernment tests all the other skills and capabilities of being a smart, connected leader covered in this book (i.e., presence, agility, collaboration, development).

3. This chapter introduced three tools: a Simple Decision-Making Model; a Simple Discernment Model; and the ABCs of Discernment, which are really 10 tools in one. In addition, this section has introduced four valuable recommendations for improving discernment efficiency and effectiveness.

Recommended Actions

1. Find out if your organization has a formal discernment process. If not, why?

2. Ask yourself how fourth industrial revolution challenges and changes will impact your organization. Discuss whether or not these changes will challenge your organization's values.

3. Have a strategic planning meeting and discuss how a discernment process could help your organization mitigate risk while also strengthening your culture, brand, and workplace inclusiveness.

Leading in the New Frontier

In the months that I researched and wrote this book, the high-velocity advancements in robotics, automation, digital transformation, artificial intelligence, and 3-D printing have considerably advanced. Every day when I log onto my favorite news feeds, there is more evidence that the future of work, the fourth industrial revolution, is here and now, and its velocity is increasing. In light of these stories of advances in AI, robotics, and automation, I hope this book has deepened your development and caused you to stop and seriously evaluate how ready your organization is for the future.

I was struck by a powerful example of how the fourth industrial revolution is impacting people's lives the day I sat down to write the first draft of this conclusion. The Federal Drug Administration (FDA) in the United States had just approved a cancer treatment, called Kymriah, for treating refractory acute lymphoblastic leukemia (ALL). This is a rare cancer that impacts hundreds of children and adolescents every year. Children diagnosed with ALL only have a 10 percent chance of living more than five years with this cancer. Prior to the FDA's approval of Kymriah, the young people and their families impacted by ALL had very little hope.

Kymriah is a fourth industrial revolution miracle. In clinical trials, 83 percent of 63 pediatric and young adult patients with ALL experienced remission. Treatment will be initially available in 32 hospitals in the United States, with a price tag of $475,000 for patients who respond favorably to the treatment. Following the approval of CTL019 (Kymriah), Scott Gottlieb, FDA commissioner, was quoted as saying, "We're entering a *new frontier* in medical innovation with the ability to *reprogram* a patient's own cells to attack a deadly cancer" [emphasis added].[1]

Indeed, we are entering a new frontier in medicine and in all aspects of society. The Kymriah case makes me think about the genetic miracle that human–machine collaboration made possible by fusing the physical, digital, and biological worlds together. But it also makes me think of the organizational miracle of people agilely coordinating their efforts, engaging in deep learning and development, and exercising mindful discernment about the safety, quality, and ethical considerations of this miracle treatment. I'm literally brought to tears when I think about the hope that young people living with ALL and their families must be experiencing right now. Fourth industrial innovation and smart, connected leadership have given them a new lease on life, a life rich with experiences that would have never been possible without a cellular "reboot."

And at the same time, I wonder, What's next? And what after that? I wonder how deep down the rabbit hole we are going, and what we *should be doing* to future-proof our organizations, communities, and global society. I don't claim to have answers to these questions, but I know that we're at an inflection point in human–technological advancement, and I know that this time it feels different than it did when the third industrial revolution really started to take root. The pace is faster and the stakes higher. This inflection point called the fourth industrial revolution is both promising and perilous, and it requires smarter, more connected leadership.

After 20 years of research and in-the-trenches experience helping leaders prepare for and implement change, I am 100 percent certain that the people who lead our governments, institutions, organizations, and businesses around the world are not ready for the journey that we're embarking on. Most lack a future-focused mindset, skills, and strategy for future-proofing their organizations. The cases of exceptional smart, connected leadership that I have seen pale in comparison to the kinds of implications that Kymriah and its next-generation offspring hold in store for us. Consequently, leaders and organizations around the world have a lot of catching up to do, and time is of the essence.

What's more, educators and those who take active roles in developing future leaders have a lot of catching up to do. The ways in which we've approached leadership development for the past 230 years has not changed much. However, over this same period of time, much has changed in the world, and it's changing faster and in new directions every day. For adult educators, coaches, consultants, HR professionals, and leaders committed to deep development, time is of the essence.

In writing this book, my objective was to answer the call of Klaus Schwab, World Economic Forum's founder and chairman, for more *responsive* and *responsible* leadership. I have called this new way of leading, "smart, connected leadership." And, as I have explained, smart, connected leadership has—at its core—five fundamental demands: presence, agility, collaboration, development, and discernment. Meeting these five demands of becoming smart, connected leaders requires what Northeastern University's president, Joseph Aoun, calls "human literacy."[2] To help refocus and deepen leaders' development around the human literacies necessary for building future-ready organizations, I've provide 25 tools and more than 20 actionable recommendations for leaders like you to put into practice right away.

The tools in this book are designed to spark conversations about how leaders need to think about and respond to 4IR challenges. These tools are also designed to deepen development and build leaders' human literacies for leading in this new frontier we call the fourth industrial revolution. But tools alone will not change organizations. Tools alone will not create future-ready organizations. For tools to be useful, they need tool users to pick them up and to take action. That's where you come in. You need to pick up these tools and practice building something special with them. There's no expectation that what you build will be perfect. But this is an invitation to start building shared futures with other people in your organization.

If you've found just one or two of these tools to be thought-provoking enough to take action on, then I have achieved my objective with this book. My challenge for you is to try out one additional tool, and to start a future-focused conversation with your peers or followers. The next challenge is for you to help broaden the conversation and start developing other leaders for the journey through the new frontier. Share ideas from this book with experienced peers. Share ideas or tools from this book with new leaders or leaders of tomorrow. Start these conversations today. Don't wait for the future to disrupt your company or institution. Share some of your personal insights from this book and join me in my mission to prepare tomorrow's leaders for the challenges ahead.

Building smarter, more connected leaders and organizations is going to require more than education. It's going to require courage and intentionality. Changing followers' mindsets about the nature of work and the mega trends impacting the world around us is a massive undertaking, and I need your help. Your followers need your help. Indeed, the world as we know it today needs your help.

If you believe that leaders can do better—if you believe that leaders must change their perspective on the way we're working, then commit to one small action. Go back through each of the chapters in this book, reviewing the chapter summaries and recommendations for action. Circle one thing you will do today to inform or inspire someone about building a shared future, and becoming a more future-ready organization.

Meanwhile, know that you're not alone. You are part of a community of smart, connected leaders. Thousands of people are having conversations about how to build future-ready organizations on Facebook, on my blog (leadership4ir.com), and in organizations and classrooms around the world. So join us. We're waiting to hear from you.

See you in the future!

Future-Proofing Strategies Tool Kit

Use the Future-Proofing Strategies Tool Kit to reference your favorite strategies in the book.

Notes

Chapter One

1. The academic and professional literature holds plenty of definitions of *leadership*. However, I like my friend and leadership expert Peter Northouse's definition: "Leadership is a process whereby an individual influences a group of individuals to achieve a common goal" (*Leadership: Theory and Practice* (6th ed.), Los Angeles, CA: Sage, 2013, p. 5).

2. Schwab, Klaus. 2017. *The fourth industrial revolution*. New York: Crown Business.

3. Brynjolfsson, Erik, and Andrew McAfee. 2014. *The second machine age: Work, progress, and prosperity in a time of brilliant technologies*. New York: Norton & Company; Ford, M. (2015). *Rise of the robots: Technology and the future threat of a jobless future*. New York: Basic Books.

4. Kellerman, Barbara. 2012. *The end of leadership*. New York: HarperCollins.

5. Schwab, Klaus. 2017. *The fourth industrial revolution* (p. 114). New York: Crown Business.

Chapter Two

1. Brynjolfsson, Erik, and Andrew McAfee. 2011. *Race against the machine: How the digital revolution is accelerating innovation, driving productivity, and irreversibly transforming employment and the economy*. Lexington, MA: Digital Frontier Press.

2. Aoun, Joseph E. 2017. *Robot-proof: Higher education in the age of artificial intelligence*. Cambridge, MA: MIT Press.

3. Taylor, Frederick. 1911. *Principles of scientific management*. New York: Harper & Brothers.

4. Frankl, Victor E. 1984. *Man's search for meaning*. New York: Washington Square Books.

5. Kahneman, Daniel. 2013. *Thinking, fast and slow.* New York: Farrar, Straus and Giroux.

6. Rock, David. "SCARF: A brain-based model for collaborating with and influencing others." *Neuroleadership Journal* 1 (2008): 1–9.

7. Rock, David. "SCARF: A brain-based model for collaborating with and influencing others." *Neuroleadership Journal* 1 (2008): 1–9.

8. Bohm, David. 2004. *On dialogue* (2nd ed.). New York: Routledge.

9. "Welcome," on Open Space (Technology) official website, accessed January 27, 2018, http://openspaceworld.org

10. Shirky, Clay. 2008. *Here comes everybody: The power of organizing without organizations.* New York: Penguin Press.

11. "How to Build a Company Where the Best Ideas Win." Ray Dalio TED Talk published on September 6, 2017 and available at www.youtube.com/watch?v =HXbsVbFAczg. Dalio discusses the Dot Collector at the 9:00-minute mark.

Chapter Three

1. Newport, Frank. 2017. "One in four workers say technology will eliminate job." *Gallup News.* Accessed May 17, from http://news.gallup.com/poll/210728 /one-four-workers-say-technology-eliminate-job.aspx.

2. Frey, Carl Benedikt, and Michael A. Osborne. 2013. "The future of employment: How susceptible are jobs to computerisation?" Oxford Martin School. Accessed January 2017, from www.oxfordmartin.ox.ac.uk/downloads /academic/The_Future_of_Employment.pdf.

3. *The future of employment: How susceptible are jobs to computerisation?* Report by Oxford Martin School and Citi. August 2017. Accessed January 2018, from www .oxfordmartin.ox.ac.uk/downloads/academic/The_Future_of_Employment.pdf.

4. Somashekhar, S. P., Martín-J. Sepúlveda, Andrew D. Norden, Amit Rauthan, Kumar Arun, Poonam Patil, Ramya Y. Ethadka, and Rohit C. Kumar. "Early experience with IBM Watson for Oncology (WFO) cognitive computing system for lung and colorectal cancer treatment." (Poster presentation American Society of Clinical Oncology, Chicago, IL, June 3, 2017). Accessed from https://meetinglibrary.asco.org/record/145389/abstract.

5. Bhalla, Vikram. *Twelve forces that will radically change how organizations work: The new way of working.* Report by the Boston Consulting Group. Accessed March 2017, from www.bcg.com/en-us/publications/2017/people-organization -strategy-twelve-forces-radically-change-organizations-work.aspx.

6. Aoun, Joseph E. 2017. *Robot-proof: Higher education in the age of artificial intelligence.* Cambridge, MA: MIT Press.

7. Setalvad, Ariha. 2015. "Demand to fill cybersecurity jobs booming." *Peninsula Press*, March 31 2015. Accessed January 2018, from http://peninsulapress .com/2015/03/31/cybersecurity-jobs-growth/.

8. Work Economic Forum, *The future of jobs: Employment, skills and workforce strategy for the fourth industrial revolution.* January 2016. Accessed November 2017, from www.weforum.org/reports/the-future-of-jobs.

9. The five future-focus mindsets I have identified through my research *are futurist, innovator, opportunist, observer, and historian.* The beta-version of the future-focus mindset assessment can be accessed via http://Leadership4iR.com /future-focus.

10. Janis, Irving L. 1971. "Groupthink." *Psychology Today* 5 (November): 43–46, 74–76.

Chapter Four

1. Deloitte Development, *Rewriting the rules for the digital age: 2017 Deloitte Human Capital Trends,* https://dupress.deloitte.com/content/dam/dup-us-en/articles /HCTrends_2017/DUP_Global-Human-capital-trends_2017.pdf.

2. Zenger, Todd. 2016. *Beyond competitive advantage: How to solve the puzzle of sustaining growth while creating value.* Boston, MA: Harvard Business Press.

3. Susskind, Richard, and Daniel Susskind. 2015. *The future of the professions: How technology will transform the work of human experts.* New York: Oxford University Press.

4. Beck, Randall, and James Harter. "Why great managers are so rare." *Harvard Business Review,* March 13, 2014, https://hbr.org/2014/03/why-good-managers -are-so-rare.

5. Hansen, M. 2009. *Collaboration: How leaders avoid the traps, create unity, and reap big results.* Harvard Business Review Press.

6. Burton, R., B. Obel, and D. D. Hakonsson. 2015. *Organizational design: A step-by-step approach.* UK: Cambridge University Press.

7. Burton, R., B. Obel, and D. D. Hakonsson. 2015. *Organizational design: A step-by-step approach* (p. 10). UK: Cambridge University Press.

8. Dutton, Jane E. 2003. *Energize your workplace: How to create and sustain high-quality connections at work.* San Francisco, CA: Jossey-Bass.

Chapter Five

1. Kegan, Robert, and Lisa Lahey. 2009. *Immunity to change: How to overcome it and unlock potential in yourself and your organization.* Boston, MA: Harvard Business School Press.

2. Bersin by Deloitte, *Transitioning to the future of work and the workplace: Embracing digital culture, tools, and approaches,* 2016, www2.deloitte.com/content/dam /Deloitte/global/Documents/HumanCapital/gx-hc-us-cons-transitioning-to-the -future-of-work-and-the-workplace.pdf.

3. According to *Rewriting the Rules for the Digital Age,* non-HR executives scored their organizations' current capabilities as adequate (33 percent), getting by (23 percent), or underperforming (14 percent), with only 25 percent rating their capabilities as good and 5 percent as excellent. Available at www2.deloitte .com/content/dam/Deloitte/global/Documents/HumanCapital/hc-2017-global -human-capital-trends-gx.pdf.

4. Bhalla, Vikram, Susanne Dyrchs, and Rainer Strack. 2017. *Twelve forces that will radically change how organizations work*, The *New* New Way of Working Series (March 27, 2017). Retrieved from www.bcg.com/en-us/publications/2017 /people-organization-strategy-twelve-forces-radically-change-organizations-work .aspx?utm_source=201704POP&utm_medium=Email&utm_campaign=otr.

5. Deloitte, *The 2016 Deloitte millennial survey: Winning over the next generation of leaders*, 2016, www2.deloitte.com/content/dam/Deloitte/global/Documents /About-Deloitte/gx-millenial-survey-2016-exec-summary.pdf.

6. Rigoni, Brandon, and Bailey Nelson. "Millennials not connecting with their company's mission." *Gallup Business Journal*, November 15, 2016, http:// news.gallup.com/businessjournal/197486/millennials-not-connecting-company -mission.aspx.

7. *2017 Global Information Security Workforce Study: Benchmarking workforce capacity and response to cyber risk*, a Frost & Sullivan Executive Briefing. Center for Cyber Safety and Education, retrieved on April 23, 2018, from https://iamcy bersafe.org/wp-content/uploads/2017/07/N-America-GISWS-Report.pdf.

8. Ford, Martin. 2015. *Rise of the robots: Technology and the threat of a jobless future*. New York: Basic Books.

9. Association for Talent Development, *2017 State of the industry,* updated February 1, 2018, www.td.org/research-reports/2017-state-of-the-industry.

10. *2015 Training industry report, Training* magazine, updated December 12, 2017, https://trainingmag.com/trgmag-article/2o15-training-industry-report.

Chapter Six

1. Ellyatt, Holly. "Nine ways WEF says it's helping to build a 'shared future.'" www.cnbc.com/2018/01/16/nine-ways-wef-says-its-helping-to-build-a -shared-future.html.

2. Bakhtin, Mikhail. 1986. *Speech genres and other late essays* (p. 7). Austin: University of Texas Press.

Conclusion

1. Herper, Matthew. "FDA approves Novartis treatment that alters patients cells to fight cancer." *Forbes* (2017, August 30), www.forbes.com/sites/matthewherper /2017/08/30/fda-approves-novartis-treatment-that-alters-patients-cells-to-fight -cancer/#260db8974400.

2. Aoun, Joseph E. 2017. *Robot-proof: Higher education in the age of artificial intelligence.* Cambridge, MA: MIT Press.

Index

About the Author

Chris R. Groscurth, PhD, PCC, is a leadership and organizational effectiveness consultant and coach. A sought-after adviser to Fortune 500, higher education, and nonprofit leaders, Dr. Groscurth has held leadership roles with Gallup Inc., the University of Michigan, and Trinity Health. He is also a strategic adviser to some of the most senior C-level executives and their teams in the world. Dr. Groscurth has published more than a dozen peer-reviewed articles, book chapters, and articles in the *Gallup Business Journal*. He received his doctorate in Human Communication Processes from the University of Georgia and has bachelor's and master's degrees in human communication studies from Western Michigan University. Dr. Groscurth is an active member of the International Coach Federation. Chris lives and works in Detroit, Michigan, and blogs at leadership4ir.com.